Lecture Notes in Mechanical Engineering

Lecture Notes in Mechanical Engineering (LNME) publishes the latest developments in Mechanical Engineering—quickly, informally and with high quality. Original research reported in proceedings and post-proceedings represents the core of LNME. Volumes published in LNME embrace all aspects, subfields and new challenges of mechanical engineering. Topics in the series include:

- Engineering Design
- Machinery and Machine Elements
- Mechanical Structures and Stress Analysis
- Automotive Engineering
- Engine Technology
- Aerospace Technology and Astronautics
- Nanotechnology and Microengineering
- Control, Robotics, Mechatronics
- MEMS
- Theoretical and Applied Mechanics
- Dynamical Systems, Control
- Fluid Mechanics
- Engineering Thermodynamics, Heat and Mass Transfer
- Manufacturing
- Precision Engineering, Instrumentation, Measurement
- Materials Engineering
- Tribology and Surface Technology

To submit a proposal or request further information, please contact the Springer Editor of your location:

China: Ms. Ella Zhang at ella.zhang@springer.com
India: Priya Vyas at priya.vyas@springer.com
Rest of Asia, Australia, New Zealand: Swati Meherishi at swati.meherishi@springer.com
All other countries: Dr. Leontina Di Cecco at Leontina.dicecco@springer.com

To submit a proposal for a monograph, please check our Springer Tracts in Mechanical Engineering at https://link.springer.com/bookseries/11693 or contact Leontina.dicecco@springer.com

Indexed by SCOPUS. All books published in the series are submitted for consideration in Web of Science.

More information about this series at https://link.springer.com/bookseries/11236

Bartosz Gapiński · Olaf Ciszak ·
Vitalii Ivanov
Editors

Advances in Manufacturing III

Volume 1 - Mechanical Engineering: Research and Technology Innovations, Industry 4.0

Springer

Editors
Bartosz Gapiński
Faculty of Mechanical Engineering
Poznan University of Technology
Poznan, Poland

Olaf Ciszak
Faculty of Mechanical Engineering
Poznan University of Technology
Poznan, Poland

Vitalii Ivanov
Department of Manufacturing Engineering,
Machines and Tools
Sumy State University
Sumy, Ukraine

ISSN 2195-4356 ISSN 2195-4364 (electronic)
Lecture Notes in Mechanical Engineering
ISBN 978-3-031-00804-7 ISBN 978-3-031-00805-4 (eBook)
https://doi.org/10.1007/978-3-031-00805-4

This Springer imprint is published by the registered company Springer Nature Switzerland AG
The registered company address is: Gewerbestrasse 11, 6330 Cham, Switzerland

Preface

This volume of Lecture Notes in Mechanical Engineering gathers selected papers presented at the 7th International Scientific-Technical Conference MANUFACTURING 2022, held in Poznan, Poland, during May 16–19, 2022. The conference was organized by the Faculty of Mechanical Engineering, Poznan University of Technology, Poland.

The aim of the conference was to present the latest achievements in the broad field of mechanical engineering and to provide an occasion for discussion and exchange of views and opinions. The conference covered topics in:

- mechanical engineering
- production engineering
- quality engineering
- measurement and control systems
- biomedical engineering.

The organizers received 165 contributions from 23 countries around the world. After a thorough peer-review process, the committee accepted 91 papers for conference proceedings prepared by 264 authors from 23 countries (acceptance rate around 55%). Extended versions of selected best papers will be published in the following journals: Management and Production Engineering Review, Bulletin of the Polish Academy of Sciences: Technical Sciences, Materials, Applied Sciences.

The book **Advances in Manufacturing III** is organized into five volumes that correspond to the main conference disciplines mentioned above.

Advances in Manufacturing III - Volume 1 - Mechanical Engineering: Research and Technology Innovations, Industry 4.0 gathers findings related to the design and manufacture of advanced modern parts and mechanisms. It covers machining, forming, and additive techniques. It presents developments concerning modern lightweight materials such as sandwich plates with auxetic core. It also discusses Industry 4.0 issues, e.g., in the field of digital twin technology or mobile robots. This book consists of 19 chapters, prepared by 56 authors from 6 countries.

We would like to thank the members of the International Program Committee for their hard work during the review process.

We acknowledge all people that contributed to the staging of MANUFACTURING 2022: authors, committees, and sponsors. Their involvement and hard work were crucial to the success of the MANUFACTURING 2022 conference.

May 2022

Bartosz Gapiński
Olaf Ciszak
Vitalii Ivanov

Contents

Response of a Sandwich Plate with Auxetic Anti-tetrachiral Core to Puncture 1
Jakub Michalski and Tomasz Strek

Comparison of the Tools for Design Process Automation in Popular CAx Systems 15
Jakub Krajanowski-Kaleta, Przemysław Zawadzki, Krzysztof Żywicki, Maciej Szymański, and Adrian Mróz

Neural Model of Manufacturing Process as a Way to Improve Predictability of Manufacturing 24
Łukasz Paśko and Katarzyna Antosz

The Role and Meaning of the Digital Twin Technology in the Process of Implementing Intelligent Collaborative Robots 39
Jakub Pizoń, Arkadiusz Gola, and Antoni Świć

Workspace Analysis of a Novel Parallel Kinematic Machine with 6 Degrees of Freedom 50
Florin Popişter, Alexandru Oarcea, Sergiu-Dan Stan, and Costan-Vlăduţ Trifan

Development of the Globelike Mobile Robot with the Manipulator Set 64
Jędrzej Urbański, Mateusz Stachowiak, and Dominik Rybarczyk

Development of Two-Wheeled Self-stabilizing Mobile Platform 78
Dawid Szatkowski and Dominik Rybarczyk

Multicriteria Optimization of the Worm Gearing Drive for Tool Magazine 93
Oleg Krol, Volodymyr Sokolov, and Aleksander Golubenko

Modification of the Spindle Head for a Milling Machine with Increased Load Capacity Drive ... 103
Oleg Krol and Volodymyr Sokolov

Theoretical Model of the Densification During Hot Pressing and its Verification ... 113
Edwin Gevorkyan, Miroslaw Rucki, Volodymyr Nerubatskyi, Zbigniew Siemiatkowski, Dmitrij Morozow, and Hanna Komarova

Fundamental Analysis of a Circular Metal Sawing Process ... 124
Dominik Brüggemann, Jens Kneifel, and Stefan Bracke

Advantages of Automatic CAM Programming in Industrial Practice – A Case Study ... 138
Maciej Kowalski and Przemysław Zawadzki

On the Machining of Aluminum Alloy Series 7 with EDM ... 149
Maria Balanou, Panagiotis Karmiris-Obratański, Nikolaos E. Karkalos, Emmanouil L. Papazoglou, and Angelos P. Markopoulos

Surface Structurization of Metallic Inserts for Enhancement of Polymer-Metal Joints ... 161
Bartosz Korzeniewski, Natalia Łożykowska, and Karol Bula

Smart Interactive Paints - The Effect of Dispergation on Color Characteristics ... 175
Zuzana Tatíčková, Michal Zoubek, Jan Kudláček, Jiří Kuchař, and Viktor Kreibich

Technology of Repairs of Anticorrosion Protection of Smaller Parts of Historic Steel Structures Using Hot-Dip Galvanizing ... 189
Jan Kudláček, Jakub Svoboda, Michal Zoubek, Henryk Kania, and Mariola Saternus

A Comparative Experimental Study of Acids Used for Anodic Oxidation for the Purposes of Aerospace Industry ... 202
Eva Michelle Sedláčková, Jiří Kuchař, Viktor Kreibich, and Jakub Horník

Analysis of Emission Solid Particles from the 3D Printing Process ... 216
Anna Karwasz and Filip Osiński

Computer Control Systems Used in Precision Agriculture ... 227
Mateusz Wiecha, Mariusz R. Rząsa, and Karolina Adamczyk

Author Index ... 239

Response of a Sandwich Plate with Auxetic Anti-tetrachiral Core to Puncture

Jakub Michalski(✉) and Tomasz Strek

Institute of Applied Mechanics, Poznan University of Technology, Poznan, Poland
jakub.ja.michalski@doctorate.put.poznan.pl

Abstract. Auxetic materials and structures have unique properties thanks to their negative Poisson's ratio resulting from specific micro- or macrostructure. They are known to have superior resistance to dynamic loads in many cases. This includes fatigue, fracture, vibrations, blasts, and impacts. However, this topic still requires further research since there are several types of auxetics and the effect of their Poisson's ratio on the response to various dynamic loads can be unexpected. In this article, impact tests were performed using the finite element method. The behavior of two sandwich plates was compared. The first plate had an auxetic anti-tetrachiral core while the other one, used for reference, was not auxetic and had a hexagonal honeycomb core. The goal was to find out whether this type of auxetic structure can be used for improved puncture protection or not. The results obtained from numerical simulations definitely confirm that the auxetic plate has the potential to replace its hexagonal honeycomb counterpart that could be used as shielding in various demanding applications.

Keywords: Auxetics · Sandwich plates · Puncture · Explicit · Dynamics · Finite element method

1 Introduction

1.1 Auxetics and Their History

Materials with a negative Poisson's ratio are called auxetics. They deform in an unusual manner when subjected to tension (they become thicker) or compression (they become thinner). This characteristic influences the behavior of auxetics also under other operating conditions. Their improved strength and resistance to various kinds of loads result in many potential applications. It has already been proven that auxetics can be used, among others, for furniture, protection equipment, sports gear, sensors and actuators, components in the aerospace industry, implants and prostheses, gaskets, or fasteners [1, 2]. This wide range of applications continues to expand with new research in the field of auxetics. It is also important to mention that auxetic structures can be made of regular materials such as structural steel or aluminum and their negative Poisson's ratio (measured as a ratio of transverse to longitudinal strain) is a result of a specific shape. The simplest example of such a structure is a re-entrant, shaped like an hourglass and obtained directly from the regular hexagonal honeycomb cell [1]. Adjustment of its internal angle

B. Gapiński et al. (Eds.): MANUFACTURING 2022, LNME, pp. 1–14, 2022.
https://doi.org/10.1007/978-3-031-00805-4_1

(between the ribs of unit cells) leads to a change in the value of Poisson's ratio. Thus, the modification of this angle allows for a smooth transition to a non-auxetic structure [3]. However, there are also many other auxetic structures, including the anti-tetrachiral one used in this case.

The history of auxetics goes back to Saint-Venant's discovery that Poisson's ratio can have negative values. After that, in the XX century, scientists confirmed negative Poisson's ratio of several natural structures (including minerals and tissues) through experiments. The large interest in this type of materials began in the 1980s when tests were extended to artificial structures [1]. The development of numerical methods allowed researchers to perform their studies on those unique materials much more efficiently. One of the greatest achievements in the field of auxetics was the invention of a method of turning regular foams into auxetic ones. This approach was first described by Lakes [4] in 1987 and then improved by other scientists. The method is based on the fact that proper mechanical and thermal treatment changes the microstructure of foams and leads to a negative Poisson's ratio.

In fact, the name auxetics was not introduced until 1991 when Evans [5] derived it from the Greek word "auxetikos" meaning "the one that tends to increase". Henceforth, materials and structures with negative Poisson's ratio are commonly referred to by this name.

Nowadays, research in the field of auxetics focuses mainly on the design of new complex micro- or even nanostructures and their advanced manufacturing methods, including precise additive manufacturing methods. Apart from that, there are still several types of loads and applications where auxetics have not been thoroughly tested yet but provide promising initial results. Some of these cases, studied by authors, are fatigue [6] and blast [7] or impact conditions. The latter are discussed in this article.

1.2 Auxetic Sandwich Plates. Blast and Impact Resistance of Auxetics

Sandwich (three-layered) plates are common composite structures used in various applications requiring high strength and low weight. This makes them particularly useful in the aerospace and defense industry. Such plates consist of two outer skins and a much thicker but porous core. The core is usually made of foam but it can be also designed as a honeycomb with a selected type of unit cell. Various auxetic sandwich panels are often tested in terms of the blast, impact, crushing, or otherwise dynamic load resistance.

The increased resistance of auxetic structures to sudden localized loads such as impact or blast can be explained based on the indentation test. This kind of experiment was carried out already by Chan and Evans [8] in 1998. The authors compared conventional and auxetic foams and noticed that the latter ones behave differently. In the case of auxetic foam, the material tends to flow towards the indenter densifying around the contact area. This results in significantly higher stiffness and lower deflection under the same load. In addition, Strek et al. [9] performed numerical simulations of contact between cylinder and plate with auxetic covering layer.

Imbalzano et al. [10] performed numerical simulations on plates with a core made of unit cells in the form of a 3D beam-like re-entrant lattice. The authors confirmed that the auxetic plate undergoes significantly lower deflection than the equivalent monolithic panel. Novak et al. [11, 12] checked the response of chiral auxetic structure used as a core in the composite plate. It was proven that auxetic core may increase resistance to blast and impact. Response of sandwich panel with chiral auxetic core to projectile impact was also verified numerically by Novak et al. [13]. Authors confirmed that the ballistic limit velocity (projectile velocity required to penetrate the structure) can be increased by introducing auxetic core.

Clearly, more previous research focuses on the blast resistance of auxetic sandwich panels and other lattice structures. Imbalzano et al. [14] compared various designs of auxetic re-entrant core with non-auxetic hexagonal honeycomb one. The authors concluded that auxetic composite panels are promising structures for applications requiring improved resistance to explosions. Al-Rifaie et al. [15] suggested that graded auxetic lattices can be used to further reinforce the blast-resistant gates. Finally, Michalski et al. [7] confirmed that sandwich plates with anti-tetrachiral core (further studied in this article) provide a significant increase in blast resistance when compared with a panel having a hexagonal honeycomb core.

Impact, crushing, and crashworthiness analyses are often performed for different types of auxetic honeycombs and other auxetic structures. Among others, Zhang et al. [16] numerically evaluated the response of re-entrant honeycomb with various cell angles subjected to dynamic crushing. Gao et al. [17] performed multi-objective crashworthiness optimization for auxetic cylindrical structure in order to increase the energy absorption.

2 Johnson-Cook Plasticity and Failure Models

2.1 Johnson-Cook Plasticity Model

Johnson-Cook plasticity model is a special form of Mises model utilizing analytical forms of hardening law and strain rate dependence. It is typically used for dynamic simulations involving the deformation of metals at high strain rates. The yield surface and flow rule in the Johnson-Cook model are the same as in the original Mises one. Isotropic hardening used in this model assumes the following relation [18–20]:

$$\sigma^0 = \left[A + B\left(\overline{\varepsilon}^{pl}\right)^n\right]\left(1 - \hat{\theta}^m\right), \tag{1}$$

where: σ^0 – static yield stress, $\overline{\varepsilon}^{pl}$ – equivalent plastic strain, A, B, n, m – measured material parameters, $\hat{\theta}$ – non-dimensional temperature which is:

- 0 for current temperature θ lower than transition temperature (at or below which no temperature dependence of the yield stress exists),
- $(\theta - \theta_{transition})/(\theta_{melt} - \theta_{transition})$ for current temperature between or equal to transition temperature or melting temperature,
- 1 for current temperature higher than melting temperature.

Optionally, strain rate dependence can be included for this model. Then the yield stress at nonzero strain rate is [20]:

$$\tilde{\sigma} = \left[A + B\left(\overline{\varepsilon}^{pl}\right)^n\right]\left[1 + C\ln\left(\frac{\dot{\overline{\varepsilon}}^{pl}}{\dot{\varepsilon}_0}\right)\right]\left(1 - \hat{\theta}^m\right), \tag{2}$$

where: $C, \dot{\varepsilon}_0$ – measured material parameters, $\dot{\overline{\varepsilon}}^{pl}$ – equivalent plastic strain rate.

2.2 Johnson-Cook Dynamic Failure Model

In Abaqus software, a dynamic failure model is available for use with the aforementioned Johnson-Cook plasticity. It is assumed that failure occurs when the sum of the quotients of the increment of the equivalent plastic strain and the strain at failure exceeds 1. The strain at failure is defined as [20]:

$$\overline{\varepsilon}_f^{pl} = \left[d_1 + d_2 exp\left(d_3\frac{p}{q}\right)\right]\left[1 + d_4 \ln\left(\frac{\dot{\overline{\varepsilon}}^{pl}}{\dot{\varepsilon}_0}\right)\right]\left(1 + d_5\hat{\theta}\right), \tag{3}$$

where: $d_1 - d_5$ – measured failure parameters, $\dot{\varepsilon}_0$– reference strain rate, p/q – pressure-deviatoric stress ratio.

3 Research Problem

3.1 Description of the Problem and Method

For this article, finite element analyses were carried out in Abaqus 2021 software. Two designs of sandwich plates were tested – auxetic and non-auxetic one. The goal was to evaluate the response of the model with auxetic core to impact load and compare it to the results of the reference simulation with the conventional plate. Explicit dynamics analyses were performed and the impactor was assumed to be rigid.

3.2 Geometry and Mesh

Both plates were modeled in such a way to keep their overall size and unit cells dimensions as close to identical as possible. However, due to the different geometry of both types of unit cells, some discrepancy was unavoidable. The same remark also applies to the relative densities of both plates. Each panel's size was about 305 × 305 × 75 mm (with two 5 mm thick skins included). The approximate size of unit cells was equal to 26 mm. Geometries of unit cells are shown in Fig. 1.

Skins were meshed with C3D8R solid elements (linear hexahedrons with reduced integration) due to their considerable thickness when compared to the thickness of cores which in turn were meshed with S4R shell elements (linear quadrilaterals with reduced integration). The mesh consisted of 22326 solid elements and 57584 shell elements (a total of 79910 elements and 82584 nodes) in the case of an auxetic plate. The conventional plate was meshed with 22326 solid elements and 58580 shell elements (a total of 80906 elements and 86954 nodes). The shell element thickness was set to 0.76 mm in both cases. All parts were positioned appropriately with respect to each other to account for shell element thickness.

In addition to panels, a rigid impactor had to be modeled. It was shaped as a hemisphere attached to a cylinder. Both the hemisphere and cylinder radius was 15 mm while the height of the cylindrical part was 30 mm. The impactor's mass was 222 g. It was meshed with 944 solid elements (C3D8R type) having a total of 1179 nodes. The impactor was placed in the middle of each plate, right above it. Meshed geometries for both analysis cases are shown in the pictures below (Fig. 2 and 3).

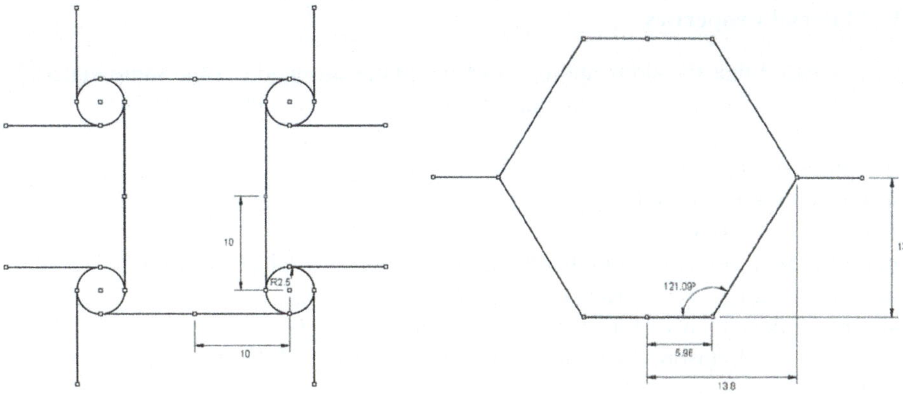

Fig. 1. Geometries of unit cells, auxetic cell on the left and non-auxetic cell on the right.

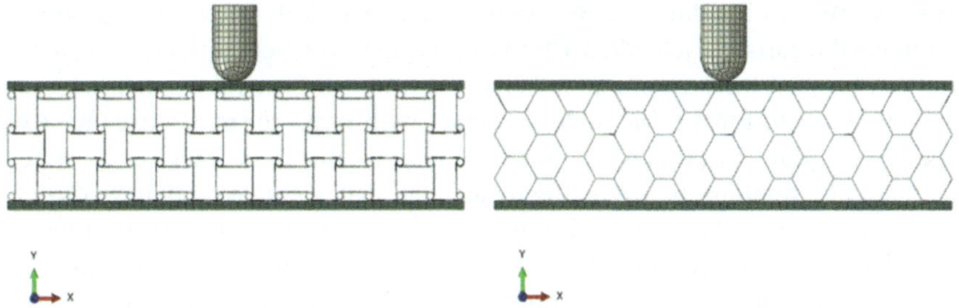

Fig. 2. Front view of meshed geometries used for both analyses, auxetic plate on the left and non-auxetic plate on the right.

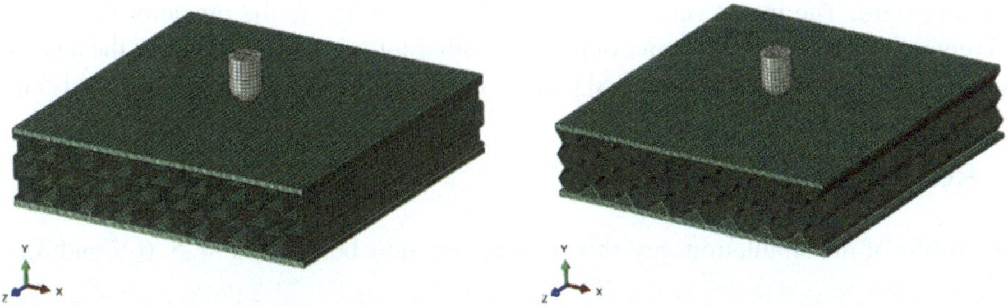

Fig. 3. Isometric view of meshed geometries used for both analyses, auxetic plate on the left and non-auxetic plate on the right.

3.3 Material Properties

It was assumed that the skins and cores of the plates are made of the same material – aluminum 6061-T6. The following material properties were defined [20, 21]:

- density: 2700 kg/m3,
- Young's modulus: 68 GPa,
- Poisson's ratio: 0.33,
- plasticity with Johnson-Cook hardening model: A = 324.1, B = 113.8, n = 0.42,
- m = 1.34, $\theta_{melting}$ = 925 K, $\theta_{transition}$ = 293.2 K,
- strain rate dependence (Johnson-Cook model): C = 0.002, $\varepsilon_0 = 1$,
- shear failure (Johnson-Cook model): $d_1 = -0.77$, $d_2 = 1.45$, $d_3 = 0.47$, $d_4 = 0$, $d_5 = 1.6$,
- specific heat: 897 J/kg*K.

3.4 Analysis Settings

The same analysis settings were used in both cases. It was assumed that skins are attached to cores via a perfect bonding, i.e. the connection cannot fail. Thus, a tie constraint was used to attach the parts to each other. This assumption is justified by manufacturing methods that can be used for such components, e.g. extrusion. General contact functionality was used to account for the frictionless contact with impactor and self-contact within the core of each plate. Friction was ignored in this case due to the fact that its effects are expected to have small influence on the results and that accurate modeling of frictional effects requires comprehensive experimental studies. In these studies, heating effects were also omitted to reduce the already high complexity of analyses. However, those effects are planned to be verified in future works, once reliable data will be obtained through experiments.

The plates were clamped on each side. The authors could have used symmetry in this case, model only one-quarter of each plate, and apply proper boundary conditions but this approach was not necessary due to sufficient computational resources and to avoid any mistakes. However, in future studies, the use of symmetry may become necessary.

Rigid body constraint was applied to the impactor so that it was assumed to be perfectly rigid. The initial velocity of 300 m/s was applied to the impactor through its reference point placed in its center of mass. Another initial condition used in the analysis was the temperature of 273 K applied to the whole model. The analysis time was 1.5 ms.

4 Results

The results of the simulations are shown in the pictures below (Fig. 4, 5, 6, 7 and 8).

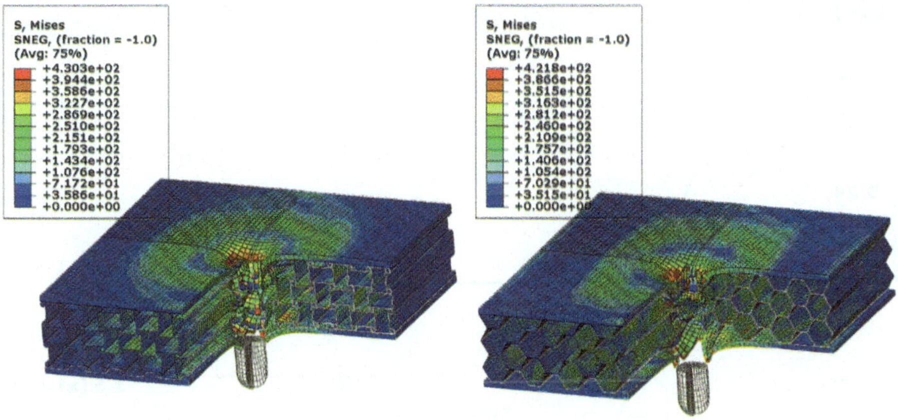

Fig. 4. Cut view of von Mises stress contour plots on the deformed shape of each model at corresponding time frames (0.625 ms), auxetic plate on the left, and non-auxetic plate on the right.

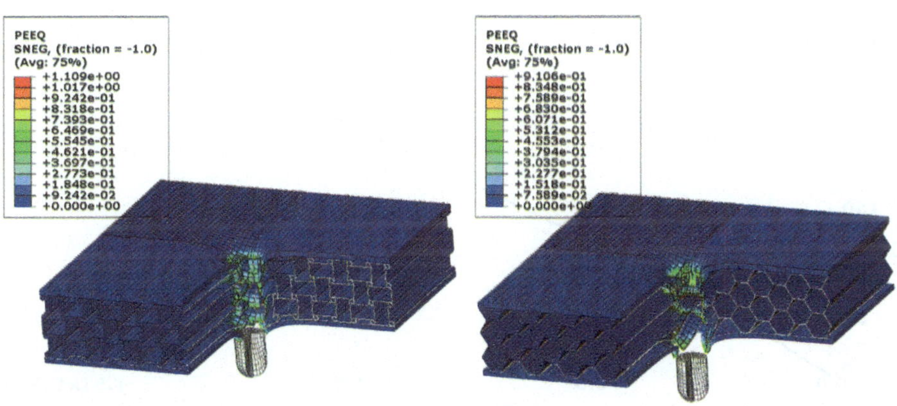

Fig. 5. Cut view of equivalent plastic strain contour plots on the deformed shape of each model at corresponding time frames (0.625 ms), auxetic plate on the left, and non-auxetic plate on the right.

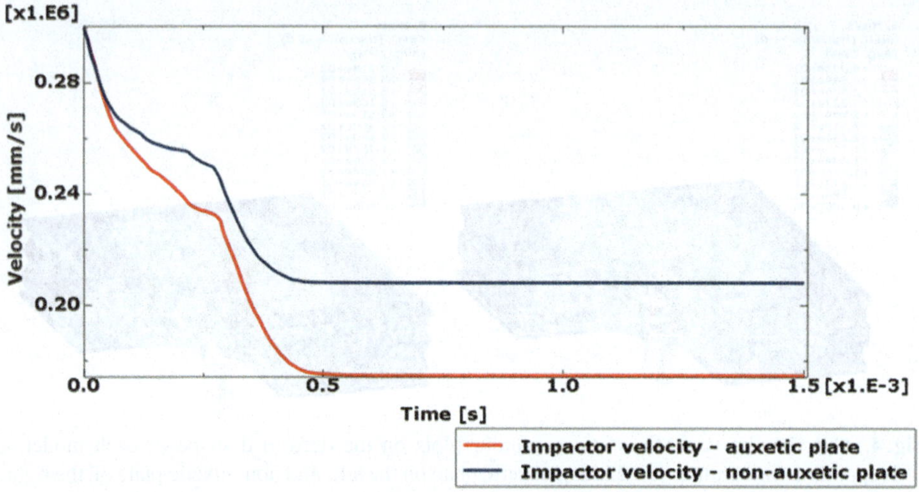

Fig. 6. Time history of impactor's vertical velocity throughout the analysis for both cases.

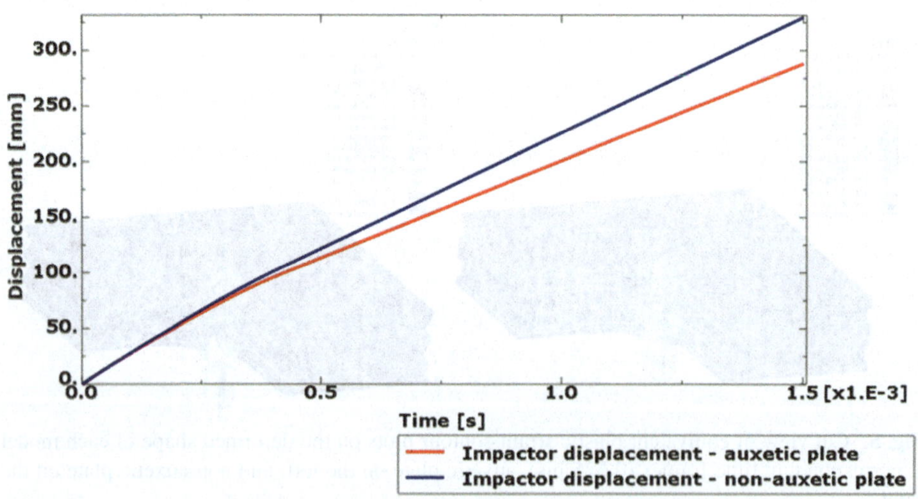

Fig. 7. Time history of impactor's vertical displacement throughout the analysis for both cases.

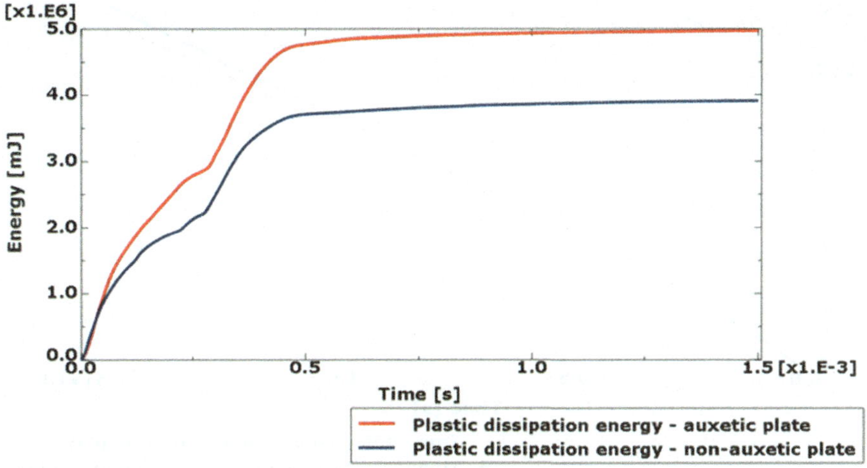

Fig. 8. Time history of plastic dissipation energy throughout the analysis for both cases.

Obtained results indicate that the auxetic anti-tetrachiral core slows down the projectile more effectively than the non-auxetic core. Stresses and plastic strains in both cases are similar but the key factor for evaluation of the performance of both plates subjected to impact is the residual velocity of the impactor. Total plastic dissipation energy was also compared for both cases to verify the energy absorption capabilities of the plates. It was found that the auxetic core increases plastic dissipation.

In order to confirm that these relationships hold regardless of the unavoidable difference in masses mentioned before, authors performed additional studies. Thickness of shell elements forming cores was adjusted in such a way that both plates had the same mass (around 1725 g). The results of this study are presented below (Fig. 9, 10 and 11).

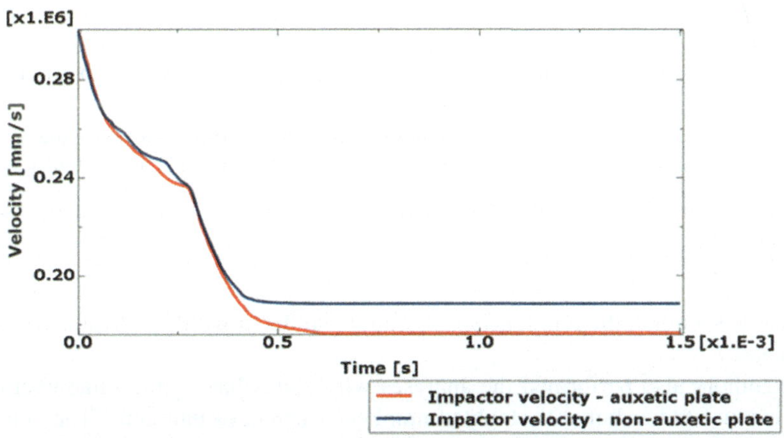

Fig. 9. Time history of impactor's vertical velocity throughout the analysis for both cases with equalized masses.

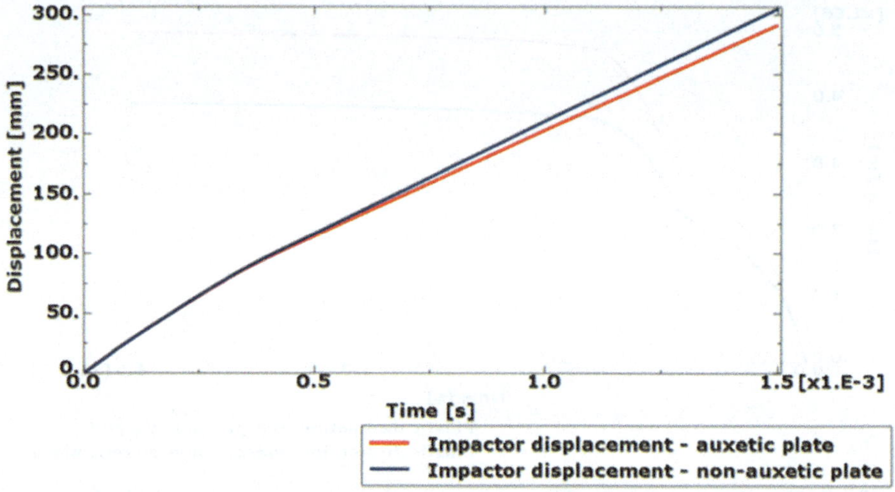

Fig. 10. Time history of impactor's vertical displacement throughout the analysis for both cases with equalized masses.

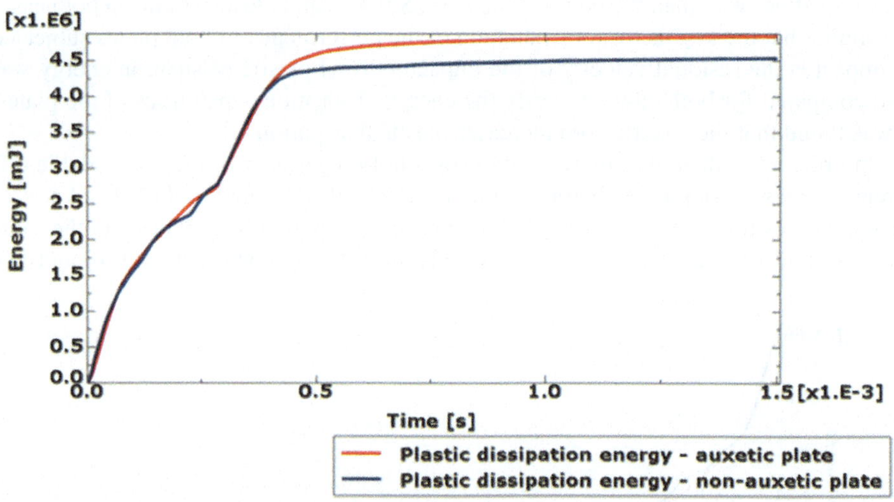

Fig. 11. Time history of plastic dissipation energy throughout the analysis for both cases with equalized masses.

It can be seen that the relationships are not invalidated with equalization of masses of the plates.

The authors also performed the analyses with plates having the same overall size as in previous studies but with doubled number of layers of unit cells. The cells were thus smaller and additional effects could be checked. The results of those analyses are presented below (Fig. 12, 13 and 14).

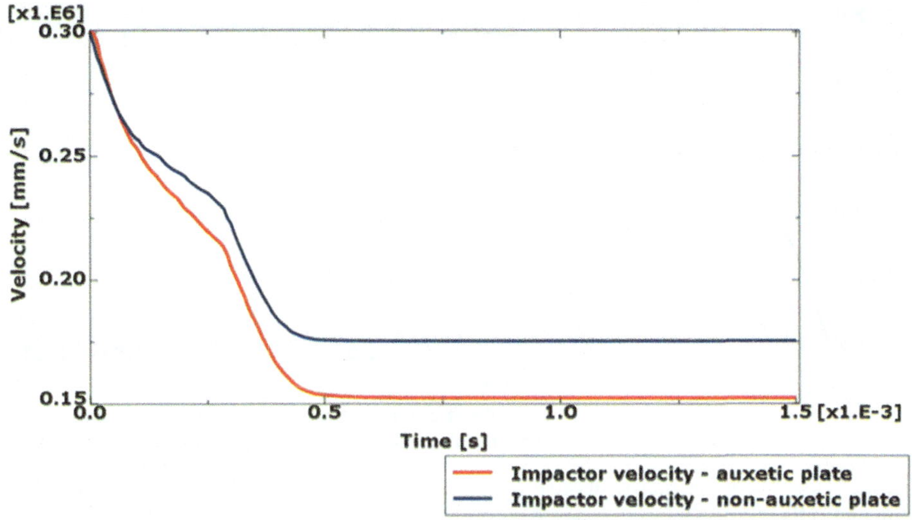

Fig. 12. Time history of impactor's vertical velocity throughout the analysis for both cases with more unit cells.

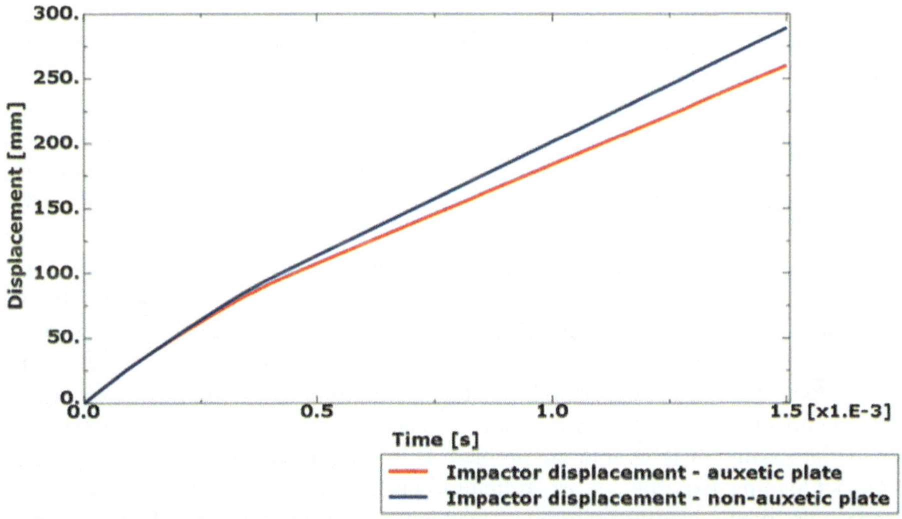

Fig. 13. Time history of impactor's vertical displacement throughout the analysis for both cases with more unit cells.

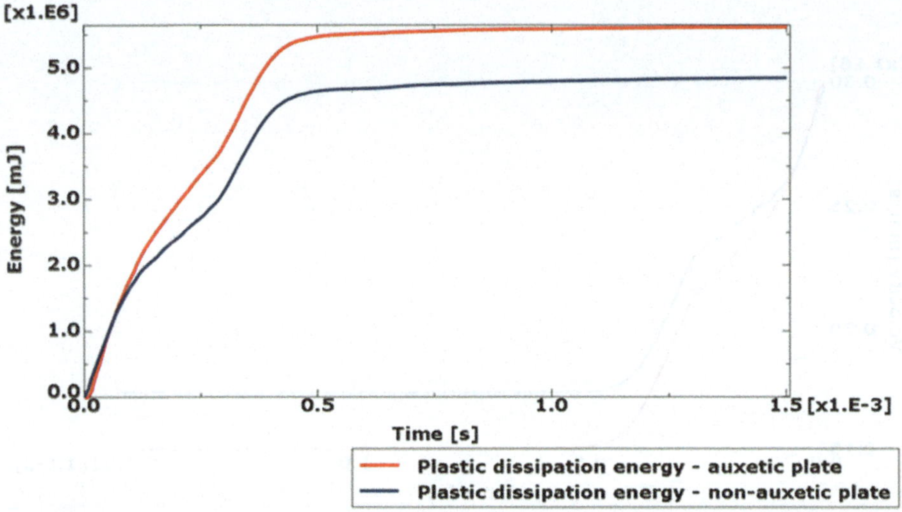

Fig. 14. Time history of plastic dissipation energy throughout the analysis for both cases with more unit cells.

The same relationships between the results can be observed also in the case of larger number of unit cells.

5 Discussion and Conclusions

Results obtained from the simulations prove that the auxetic anti-tetrachiral core performs better in terms of response to puncture than the non-auxetic hexagonal honeycomb core.

However, it is important to mention that various simplifications were used in the numerical model. There are several sources of potential errors. Mesh density was carefully selected to avoid significant dependency of the results on the level of discretization. In this case, the material model was crucial for accurate output. Because of that, material data had to be taken from reliable source. Another approximation used in the simulations is the connection between skins and cores of both plates. It was assumed that this connection cannot break during the analysis. To model it more realistically, it would be necessary to account for the potential delamination and failure of the connection. But such an approach would require additional data regarding the strength of the adhesive.

Comparison of the numerical results to data acquired from physical tests could provide interesting insights into the problem being considered here. However, in cases of impact loading, experiments are difficult to perform and potentially dangerous. This is especially true for significant projectile velocities, as is the case here. Thus, numerical methods are often used when evaluating the impact resistance of various structures.

Results obtained for the purpose of this article confirm that, in analyzed conditions, auxetic anti-tetrachiral core can be used in place of the regular hexagonal honeycomb structure to increase the puncture protection capabilities of sandwich plates. Such panels could be used in various protective structures, among others for the aerospace and defense industry where lightweight honeycomb structures are already used but may still require further strengthening against high-velocity impacts.

Acknowledgments. This work was supported by grant of the Ministry of Science and Higher Education in Poland: 0612/SBAD/3576 (2021/2). Simulations were carried out during an internship in BudSoft company.

References

1. Lim, T.C.: Auxetic Materials and Structures. Springer, Singapore (2015)
2. Ren, X., Das, R., Tran, P., Ngo, T.D., Xie, Y.M., Auxetic metamaterials and structures: a review. Smart Mater. Struct. **27**(2), 023001 (2018)
3. Bilski, M., Piglowski, P.M., Wojciechowski, K.W.: Extreme Poisson's ratios of honeycomb, re-entrant, and zig-zag crystals of binary hard discs. Symmetry **13**(7), 1127 (2021)
4. Lakes, R.: Foam structures with negative Poisson's ratio. Science **235**(4792), 1038–1040 (1987)
5. Evans, K.E.: Auxetic polymers: a new range of materials. Endeavour **15**(4), 170–174 (1991)
6. Michalski, J., Strek, T.: Fatigue life of auxetic re-entrant honeycomb structure. In: Gapiński, B., Szostak, M., Ivanov, V. (eds.) MANUFACTURING 2019. LNME, pp. 50–60. Springer, Cham (2019). https://doi.org/10.1007/978-3-030-16943-5_5
7. Michalski, J., Strek, T.: Blast resistance of sandwich plate with auxetic anti-tetrachiral core. Vib. Phys. Syst. **31**(3), 20317 (2020)
8. Chan, N., Evans, K.E.: Indentation resilience of conventional and auxetic foams. J. Cell. Plast. **34**(3), 231–260 (1998)
9. Strek, T., Matuszewska, A., Jopek, H.: Finite element analysis of the influence of the covering auxetic layer of plate on the contact pressure. Phys. Status Solidi B **254**(12), 1700103 (2017)
10. Imbalzano, G., Tran, P., Ngo, T.D., Lee, P.V.S.: Three-dimensional modelling of auxetic sandwich panels for localized impact resistance. J. Sandwich Struct. Mater. **19**(3), 291–316 (2017)
11. Novak, N., Starcevic, L., Vesenjak, M., Ren, Z.: Blast and ballistic loading study of auxetic composite sandwich panels with LS-DYNA. In: 12th European LS-DYNA Conference 2019, Koblenz, Germany (2019)
12. Novak, N., Starcevic, L., Vesenjak, M., Ren, Z.: Blast response study of the sandwich composite panels with 3D chiral auxetic core. Compos. Struct. **210**, 167–178 (2019)
13. Novak, N., Vesenjak, M., Kennedy, G., Thadhani, N., Ren, Z.: Response of chiral auxetic composite sandwich panel to fragment simulating projectile impact. Phys. Status Solidi B **257**(10), 1900099 (2020)
14. Imbalzano, G., Linforth, S., Ngo, T.D., Lee, P.V.S., Tran, P.: Blast resistance of auxetic and honeycomb sandwich panels: comparisons and parametric designs. Compos. Struct. **183**, 242–261 (2018)
15. Al-Rifaie, H., Sumelka, W.: Improving the blast resistance of large steel gates–numerical study. Materials **13**(9), 2121 (2020)
16. Zhang, X., Ding, H., An, L., Wang, X.: Numerical investigation on dynamic crushing behavior of auxetic honeycombs with various cell-wall angles. Adv. Mech. Eng. (2015)
17. Gao, Q., Zhao, X., Wang, C., Wang, L., Ma, Z.: Multi-objective crashworthiness optimization for an auxetic cylindrical structure under axial impact loading. Mater. Des. **143**, 120–130 (2018)

18. Stopel, M.: Methodology for determining constants for Johnson-Cook constitutive and damage model. University of Sciences and Technology in Bydgoszcz, Bydgoszcz (2020). (in Polish)
19. Jankowiak, T.: The use of experimental methods and computer simulation for determining the properties of materials subjected to high deformation rate. Poznan University of Technology, Poznan (2016). (in Polish)
20. Dassault Systemes SIMULIA Abaqus: 2021 software documentation
21. Dassault Systemes SIMULIA Abaqus Technology Brief: Simulation of the ballistic perforation of aluminum plates with Abaqus/Explicit

Comparison of the Tools for Design Process Automation in Popular CAx Systems

Jakub Krajanowski-Kaleta[1], Przemysław Zawadzki[2(✉)], Krzysztof Żywicki[2], Maciej Szymański[1], and Adrian Mróz[1,3]

[1] STER Institute Maciej Szymanski, Ster 1 Street, 3, 62-080 Swadzim, Poland
[2] Faculty of Mechanical Engineering, Poznan University of Technology, Piotrowo 3, 60-965 Poznań, Poland
przemyslaw.zawadzki@put.poznan.pl
[3] Calisia University, Poznańska 201-205 Street, 62-800 Kalisz, Poland

Abstract. The paper presents a comparison of the use of two popular CAx systems in the context of the automation of the chosen design tasks. The presented solution (automation of the design of seat mounting) is part of a larger system, intended for support and management of public transport vehicles seat production. Based on the accumulated knowledge and experience, regulations, and manufacturing guidelines provided by STER Company, a new design procedure was developed and tested separately in the 3DExperiecne Catia program and Siemens NX software. The results of verification were presented and discussed showing common features, advantages, disadvantages, and limitations in the context of using generative CAD models.

Keywords: Design automation · Generative CAD modeling · KBE systems

1 Introduction

Although it has been assumed that each stage in the product development cycle is important and may affect the final success of the product on the market, from a practical point of view, the key to this success seems to be good design. The design process is very often the most important and labor-intensive stage [1]. Mistakes made at this stage adversely affect not only the production processes but also often cause failure on the market. Considering such factors like analyze the customer expectations, determining the technical and functional requirements of the new product, and the limitations resulting from the applied manufacturing processes, it can be said that designers have the greatest responsibility in implementing new solutions indeed.

In the area of development and optimization of production processes, much attention is paid to modern, intelligent technologies referred to as Industry 4.0 [2]. It should be noticed that this idea is really a way to increase the flexibility of production of enterprises that are to be ready to meet the individual requirements of their clients [3]. This is closely related to the concept of mass customization strategy, which was introduced in theoretical considerations several decades ago. Industry 4.0 is really a way to effectively

B. Gapiński et al. (Eds.): MANUFACTURING 2022, LNME, pp. 15–23, 2022.
https://doi.org/10.1007/978-3-031-00805-4_2

implement this idea in practice today. One of the basic issues of mass customization is ensuring an economically acceptable unit cost of the product. To make personalized products competitive, they should not be much more expensive than those offered in mass production [4]. Customers do not want to pay for personalization in themselves, they are ready to pay only for the additional value that the product has for them.

Regarding the above and focusing on the area of design a question could be asked: how designers should design faster (so that it was cheaper) and more accurately (here considering the individual approach to customer requirements), maintaining appropriate quality standards (i.e. minimizing mistakes) in the same time? The answer to this question could be - the generative design method, using parametric CAD systems.

Today's CAx class systems allow the use of such solutions, thanks to which the designers' work is supported by various tools, such as parts and connection generators, tools that allow automating the construction of a series of parts, the selection of materials, shape optimization in terms of strength, ergonomics, and recycling [5–8]. In some systems, this type of solution is offered as standard, while in others a package of appropriate functions allows you to develop them yourself. In more complex or specialized cases of designing, developing an expert system usually requires many hours of preparatory work, collection, and recording of engineering knowledge, programming work, and, above all, testing before this type of solution is ready for implementation.

In this paper, the authors present a comparison of the use of two popular CAx systems in the context of the automation of the chosen design tasks. Presented solution (automation of design of seats mounting) will be a part of a larger system, intended for support and management of public transport vehicles seat production. Based on the accumulated knowledge and experience, regulations, and manufacturing guidelines provided by STER Company, a new design procedure was developed and tested separately in the 3DExperiecne Catia program and Siemens NX software.

2 Generative CAD Modelling

Generative CAD modeling is of particular importance for enterprises dealing with mass production of products, that are similar but are of diverse characteristics at the same time. An example of such a process can be the production of automotive parts, where very similar components, having the same purpose, may differ from each other due to different variants of the vehicle [4]. It is similar in many other industries, such as household appliances, sanitary, architecture or consumer electronics [9–11].

The idea of generative design is to reduce the time to develop a new product variant by automating repetitive activities in the design process. This automation is provided by special CAD models that allow you to quickly generate a new product variant in accordance with applicable requirements. Generative design is, in fact, a special approach to design, where the designer does not focus on a specific case (product variant), but on all the features describing the whole family of similar products. Using appropriate tools of parametric CAD programs, the scope of product family configuration variables should be described - which features can be changed and in what range, and what is more difficult – describe how those features are related to each other. Although in practice it can be very difficult to develop these assumptions (it is necessary to predict what the customer will want to adopt in a given product), it is worth doing because it is estimated that approx. 80% of design time concerns repetitive, routine tasks [12, 13]. Their acceleration can therefore significantly affect the optimization of the entire product development cycle and savings.

There are many examples of the use of the generative approach in design process in the literature. Wu et al. [14] introduced a method for the generative design of face gears, where they proposed non-equidistant isoparametric discretization, so the tooth surface points are generated as an approximately even distribution on the tooth surface. Żywicki et al. [6] in their works described, among others implementation of generative CAD models in the model of production processes simulated in the Smart Factory Laboratory. At work Gorski et al. [15] the authors presented the use of a generative CAD model for design bus interior, as well as a way of data exchange between the CAD program and the VR system. LE et al. [16] described an innovative Product Development solution using parametric design and Generative modeling. In the work Grabowik et al. [17] one can learn about the knowledge-based design method using the NX Siemens system.

3 Research Methodology

3.1 Case Study – Smart Production of Public Transport Vehicle Seats

The public transport vehicle (ex. bus, trains) production process is quite complicated because unlike typical automotive production, where the number of cars variants is well known and rather limited, public transport vehicles can be produced as one-off production. In addition, the large number of features that affect the final form of the vehicle means that the number of possible variants is hypothetically unlimited. This also has an impact on described example, because the seat manufacturer must be ready to develop a new mounting shape variant for each order (sometimes it is also necessary to create a new seat variant, however in this paper authors are limited only to seat mounting design).

To meet nowadays expectations of the market STER Company is developing a new strategy, according to the idea of a smart production model (Industry 4.0). One of the main assumptions of this transformation is to create intelligent production and management system, which allows to gather clients' demands and translate them into productions systems requirements automatically (or semi-automatically in some steps). The main parts of the system are:

- superior system of data processing;
- identify clients expectations by special Configurator software – the creation of order data;
- automatically import and translate the order data to CAD software;
- use generative CAD models to speed up the design process;
- translate the technical requirements of the new design and assign them to technological process;
- management of production process according to generated order data.

For this paper, only one step of the whole system is being analyzed, related to the use of generative CAD models. The main aim was to check two different software in such CAD models preparation process.

3.2 Design Process Requirements

The flexibility of seat configuration on the public transport vehicle (ex. bus) is therefore a big problem for the seat manufacturer's engineers. They must prepare for each new order the appropriate technical documentation of the seat mounting, considering the technical and approval conditions of the seat distribution (mounting methods, distance between seats, etc.). There are also a few different types of mounting (Fig. 1). In addition, the source data describing the context of the new project (vehicle manufacturer's documentation) vary from customer to customer. So, it could be a 3D models or more often 2D documentation showing the interior of the vehicle. This environment must be considered in the mounting design process, as structural components (such as fastening strips) and vehicle equipment (e.g. heaters) are important. In practice, there are known cases where for each seating position (at least several dozen) there was a different mounting.

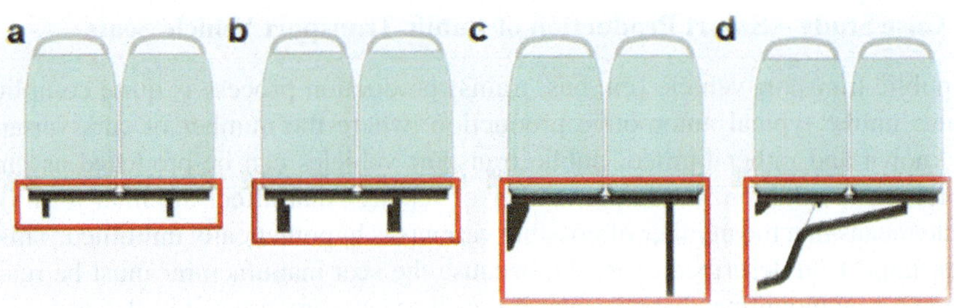

Fig. 1. Different types of seats mounting: (a) "leg – leg"; (b) "mudguard"; (c) "leg – wall"; (d) "cantilever"

Due to the above factors, the development of technical documentation (which is needed in further processes), is laborious (it can take a few days for one order carried out by one designer). In addition, although the attachment consists of an average of a dozen or so elements, engineers have difficulty finding suitable archival solutions (there is too much to search effectively). That is why they often duplicate their work by creating new adapters or their parts without knowing if there are already developed matching solutions. This is why the generative CAD model for the intelligent design process is justified.

3.3 Research Plan

Automation in 3DExperiecne Catia can be done in some different ways. The most popular option is to use a toolkit called "Knowledgeware" containing, inter alia, embedded application "Engineering Rules Capture". This application uses EKL (Enterprise Knowledge Language) to create scripts embodied in the model tree. The language is easy to use and fast. It is designed to create applications and perform their broadly understood integration. This method also allows the use of Visual Basic, which is much slower and more difficult to use, but is more versatile and has a wider application, so it can be used as a fallback solution easily. Engineering Rules Capture is generally an option with limited capabilities, but gives very quick results due to ease of use. Also, its products are very reliable and fast, but it requires the purchase of an additional license. This method was used in the experiment described in this publication.

The second method of automating Catia is scripting in Visual Basic, Visual Basic for Applications, or CATScript. This allows the use of a wide range of functions from many Catia modules. Catia allows you to record activities carried out in the graphical environment and saves them as ready-made macros that can be edited or used in the future. The solution is suitable for small projects. Code development is much more time-consuming than the previous method, and the code is slower to execute.

Automation in Siemens NX can be done in a number of ways, which differs significantly from those used in Catia. The first is NX Open. It is an API - Application Programming Interface - set of functions that can be used in a custom program and its integration with third-party applications. NX Open use allows for the automatic creation of parts, assemblies and other objects in accordance with the developed algorithm It is intended mainly for large projects. The program can be written in 5 popular languages: VB, C#, C++, Java and Python, which is convenient for programmers. A feature of this method is its level of advancement, which requires a lot of knowledge from the programmer to start working. There is also a time-consuming configuration of the external environment. The code written by the programmer must be compiled and made available as a separate library, and not, as in the case of Catia, entered into the structure of the CAD model tree. As above, this solution also requires a license. IDE used in the presented project is Microsoft Visual Studio.

Other methods of programming in NX are limited to simple, usually single-file scripts, ensuring the automation of a single action or creation. You can use SNAP (Simple NX Application), GRIP (Graphics Interactive Programming), Knowledge Fusion (interpreted, object-oriented language) or Manuscript. The first of these is API for simple geometry creating. SNAP represents a higher abstraction layer than NX Open and it has lower possibilities. It was designed to make automation as easy as it could be to target average NX user. Figure 2 shows the layer diagram. The lowest and therefore the fastest and most reliable are NX internal functions. These are all features available throughout this CAD software. User functions has lower functionality as it's an API written in C. NX Open provides full functionality but is very complicated. SNAP is much simpler and does not provide some features. The next method – GRIP – is just an old version of SNAP provided to keep backward compatibility, but it is not developed anymore. The Knowledge Fusion is a procedural language used to capture progress in KBE applications. Finally, manuscript is used for Customize menus and NX functions.

Automation analysis in Catia and NX systems consisted in making a similar project in both systems and then comparing their advantages and disadvantages, both from the practical and theoretical side. The time of development and execution of the project code is the objective result of the test. The comparison also took place on other planes, which gave a complete, comprehensive view that allows us to get to know the features of each

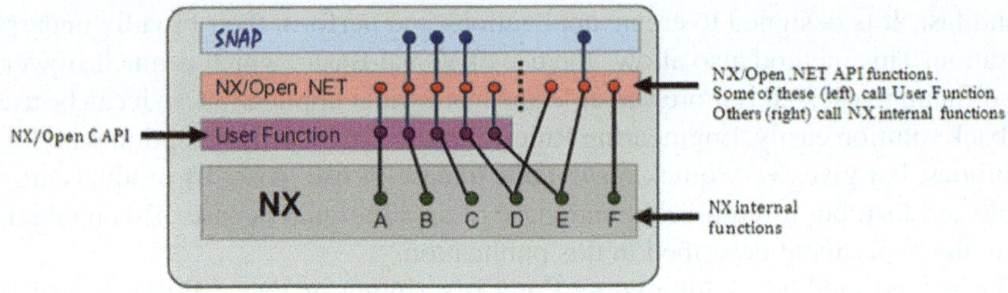

Fig. 2. NX automation layers [18]

of the systems mentioned. Automation was prepared using the best tools available in a given software. The project designed to compare the 3DX Catia system with Siemens NX is based on the automatic spacing of the seat models in the bus model. In both systems, an analogous script was prepared that performs the following activities:

- looks for a suitable armchair model in the repository;
- loads a table with target seat positions;
- sets the seat models in appropriate positions in relation to the bus geometry (Fig. 3).

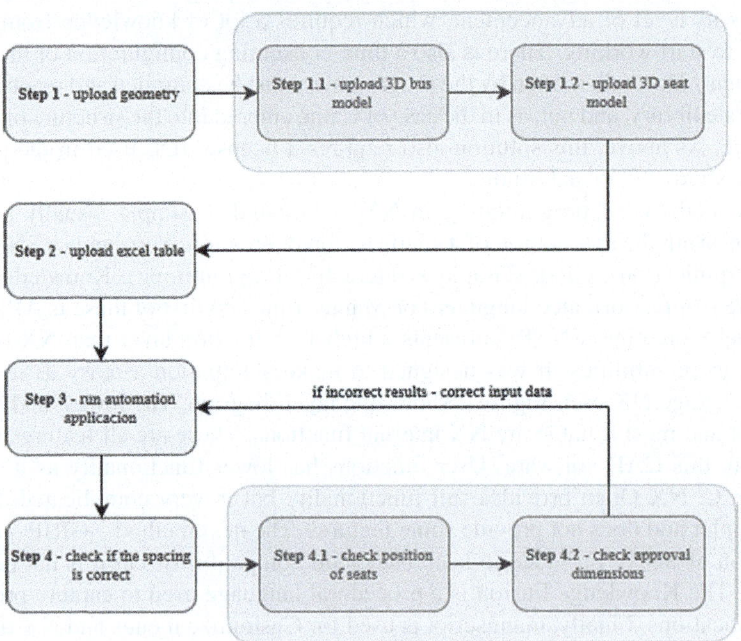

Fig. 3. Application workflow

This automation significantly speeds up the operation of setting the armchairs to the appropriate positions, because the code execution takes less than a minute, and the user would spend several minutes on it by manually adjusting them (Fig. 4).

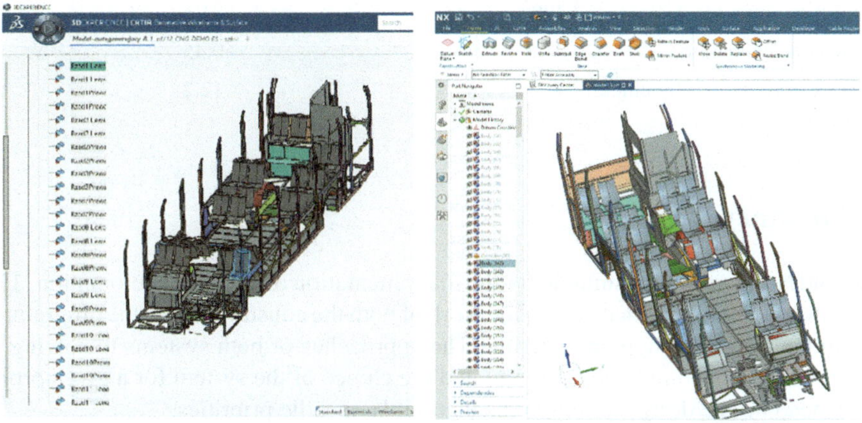

Fig. 4. The result of the developed program in Catia and NX systems

4 Results

The result of comparing CAD systems is presented in Table 1. The most important parameters are: the time of creating the program by writing the automation code and the time of executing this automation by the computer. The first parameter, measured in man-hours, describes the amount of work a programmer puts in to achieve certain results. It was assumed that the programmer had the same experience in both tested systems and a completely prepared automation environment. The time counted from the start of work to the first successful automation project. This parameter determines, among others ease of use of a given CAD system, availability of functions and their universality. The level of the programming language has a direct impact on the development time of the solution. Specifies an abstraction level the programmer uses when writing the statement. As a rule, the higher this level, the faster the programming is, however, it is less universal. The lower level means more freedom for the programmer due to the less structured and schematized functions available. However, it then requires writing more code.

The other important parameter determined the computing efficiency of the CAD system. Faster execution of the automation code compared to the competing system means its better performance, which translates directly into the profitability of working with it for more complex tasks. While the automation code is being executed, the system is frozen so that further work with it may not take place until the automatic operation is completed.

Table 1. Comparison of the time of performed tasks

Task	3D Experience Catia Time [min]	Siemens NX Time [min]
Code development	408	600
Script development	408	840
Single code execution	0.18	0.45
Code executions for all vehicle positions	5	12.2

5 Summary

Catia and NX are systems suitable for the implementation of advanced automation. They have extensive solutions that help in the work of both the constructor and the programmer who automates the design engineering. The approaches of both systems to solving the same problem are completely different and the choice of the system for a given project will depend on the design assumptions that imply specific priorities.

The comparison was settled in favor of Catia, but the choice of software will be more appropriate if it is preceded by a meticulous analysis of the projects that will be implemented in it. Each of the CAD software also has its own limitations, the analysis of which helps to avoid problems during work. There are activities that cannot be automated in one software, while in the other it is performed without any problems.

The programmer's experience should also be considered. If he prefers to focus on simple activities that give quick results, he should choose Catia. If you are familiar with development environments and languages supported by NX, this software may be the right choice.

Acknowledgements. Research curried out in the frame of Project co-financed by The National Center for Research and Development under the European Regional Development Fund program under the Intelligent Development Program. The project No POIR.01.01.01-00-0480/18 is carried out as part of the of The National Center for Research and Development Program and another, co-financed by Wielkopolska Region under the European Regional Development Fund program under Regional Operational Program for Wielkopolskie Voivodeship - the project No RPWP.01.02.00-30-0014/16.

[PL] Projekt współfinansowany przez Narodowe Centrum Badań i Rozwoju w ramach programu Europejskiego Funduszu Rozwoju Regionalnego w ramach Programu Inteligentny Rozwój. Projekt realizowany w ramach konkursu Narodowego Centrum Badań i Rozwoju: Szybka Ścieżka nr POIR.01.01.01-00-0480/18 oraz innego, współfinansowanego przez Samorząd Województwa Wielkopolskiego w ramach Programu Operacyjnego Województwa Wielkopolskiego - projekt nr RPWP.01.02.00-30-0014/16.

References

1. Górski, F., Buń, P., Wichniarek, R., Zawadzki, P., Hamrol, A.: Design and implementation of a complex virtual reality system for product design with active participation of end user. In: Amaba, B. (eds.) Advances in Human Factors, Software, and Systems Engineering. AISC, vol. 492, pp. 31–43. Springer, Cham (2016). https://doi.org/10.1007/978-3-319-41935-0_4
2. Brettel, M., Friederichsen, N., Keller, M., Rosenberg, M.: How virtualization, decentralization and network building change the manufacturing landscape: an industry 4.0 perspective. Int. J. Mech. Aerosp. Ind. Mechatron. Manuf. Eng. **8**(1) (2014)
3. Hamrol, A., Gawlik, J., Sładek, J.: Mechanical engineering in industry 4.0. Manag. Prod. Eng. Rev. **10**, 14–28 (2019)
4. MacCarthy, B., Brabazon, P.G., Bramham, J.: Fundamental modes of operation for mass customization. Int. J. Prod. Econ. **85**(3), 289–304 (2003)
5. Górski, F., Zawadzki, P., Hamrol, A.: Knowledge based engineering as a condition of effective mass production of configurable products by design automation. J. Mach. Eng. **16**(4), 5–30 (2016)
6. Żywicki K., Zawadzki P., Hamrol A.: Preparation and production control in smart factory model. In: Rocha, Á., Correia, A., Adeli, H., Reis, L., Costanzo, S. (eds.) Recent Advances in Information Systems and Technologies. WorldCIST 2017. AISC, vol. 571, pp.519–527. Springer, Cham. https://doi.org/10.1007/978-3-319-56541-5_53
7. Dostatni, E., Diakun, J., Grajewski, D., Wichniarek, R., Karwasz, A.: Functionality assessment of ecodesign support system. Manag. Prod. Eng. Rev. **6** (1), 10–15 (2015)
8. Reddy, E.J., Sridhar, C.N.V., Rangadu, V.P.: Knowledge based engineering: notion, approaches and future trends. Am. J. Intell. Syst. **5**(1), 1–17 (2015)
9. Fogliatto, F.S., Da Silveira, G.J., Borenstein, D.: The mass customization decade: an updated review of the literature. Int. J. Prod. Econ. **138**(1), 14–25 (2012)
10. Salvador, F., De Holan, P.M., Piller, F.: Cracking the code of mass customization. MIT Sloan Manag. Rev. **50**(3),71–78 (2009)
11. Krish, S.: A practical generative design method. Comput. Aided Des. **43**(1),88–100 (2011)
12. Górski, F., Hamrol, A., Kowalski, M., Paszkiewicz, R., Zawadzki, P.: An automatic system for 3D models and technology process design. Trans. FAMENA **35**(2), 69–78 (2011)
13. Chandrasegaran, S., et al.: The evolution, challenges, and future of knowledge representation in product design systems. Comput. Aided Des. **45**(2), 204–228 (2013)
14. Wu, Y., Zhou, Y., Zhou, Z., Tang, J., Ouyang, H.: An advanced CAD/CAE integration method for the generative design of face gears. Adv. Eng. Softw. 126, 90–99 (2018)
15. Gorski, F., Zawadzki, P., Bun, P., Wichniarek, R., Szalanska, K.: Virtual reality and CAD systems integration for quick product variant design. In: Hamrol, A., Ciszak, O., Legutko, S., Jurczyk, M. (eds.) Advances in Manufacturing. LNME, pp. 599–608. Springer, Cham (2018). https://doi.org/10.1007/978-3-319-68619-6_57
16. Chi Hieu, L.E., et al.: Customer driven mass-customisation and innovative product development with parametric design and generative modelling. In: Gao, J., El Souri, M., Keates, S. (eds.) Proceeding of the 15th International Conference on Manufacturing Research. Series: Advances in Transdisciplinary Engineering XXXI, vol. 6. IOS Press (2017)
17. Grabowik, C., Kalinowski, K., Kempa, W., Paprocka, I.: A method of computer aided design with self-generative models in NX Siemens environment. In: IOP Conference Series: Materials Science and Engineering, vol. 95, no.1, p. 012123. IOP Publishing (2015)
18. Getting Started with NXOpen–online materials. https://docs.sw.siemens.com/en-US/product/209349590/doc/PL20200522120320484.gs_nx_open/pdf/

Neural Model of Manufacturing Process as a Way to Improve Predictability of Manufacturing

Łukasz Paśko(✉) and Katarzyna Antosz

Faculty of Mechanical Engineering and Aeronautics, Rzeszów University of Technology, Powstańców Warszawy 8, 35-959 Rzeszów, Poland
{lpasko,katarzyna.antosz}@prz.edu.pl

Abstract. The aim of this paper is to present the concept of a neural model of a manufacturing process. The goal of the model is to forecast the number of defective products based on the historical values of the manufacturing process parameters and the historical values of the number of defective products. The paper describes the creation of the model on the example of a manufacturing process in a glass factory. The use of the model allowed for a more accurate prediction of the number of defective products, which is treated as the improvement of the manufacturing predictability mentioned in the title of the paper. The model includes several NARX artificial neural networks. Each NARX network considers data from a different part of the production line. The forecast results on four test data sets are also presented. These results were compared with the classic approach, which uses a single neural network. The created model allowed for a significant reduction in the prediction error in four test data sets considered.

Keywords: Quality prediction · Neural networks · Glass factory · Defective products · Manufacturing data

1 Introduction

The production process is defined as the process of transforming the input vector of the production system into the output vector. The input vector consists of technical means of production, work items, energy factors, personnel, capital, and information. In turn, the output vector includes products, services, information, waste, as well as defective products. Predicting the results of a manufacturing process that are part of the output vector can be a difficult task, especially when dealing with complex production processes. Therefore, one of the most important tasks of modern production engineering is the development of methods and techniques allowing to predict what will be the result of the production process. Exemplary predictive tasks in this context may be related to the prediction of the number of products produced at the given values of the input vector, as well as the prediction of defective products produced at the given values of the manufacturing process parameters.

B. Gapiński et al. (Eds.): MANUFACTURING 2022, LNME, pp. 24–38, 2022.
https://doi.org/10.1007/978-3-031-00805-4_3

The elements of the output vector, the prediction of which can be particularly important, include the number of non-compliant or defective products. The task of predicting the number of defects is a part of quality assurance. Product quality assurance is a key element of the strategy of modern manufacturing companies. Therefore, the role of quality assurance research is becoming more and more significant. Such research can help identify the causes of defects in products or determine the method of eliminating product defects.

Information extracted from data that comes from a manufacturing process can be used to support the quality assurance process. Currently, many manufacturing companies face the problem of huge amounts of data. These data are generated, inter alia, by various types of devices that carry out operations in the technological process or also by production process monitoring systems. Properly conducted analysis of production data can reveal important information useful for predicting the quality of products. In the era of rapid development of technologies for collecting and processing large amounts of data, it is worth paying special attention to the possibility of supporting the quality assurance process based on data obtained from the manufacturing process.

This paper proposes an approach using the exploration of production data and machine learning techniques in the problem of predicting the number of defective products. A set of artificial neural networks (ANNs) was used for prediction. Each ANN included in the set is a NARX network (nonlinear autoregressive with external input). The proposed approach allows to reduce the prediction error compared to a single NARX network. It was demonstrated on an example of data obtained from the production process in a glass factory. The set of ANNs constitutes a model of a part of the manufacturing process from which the obtained data are derived.

The rest of the paper is organized as follows. Section 2 describes the Cross-Industry Standard Process for Data Mining (CRISP-DM) that was used in the methodology introduced in this paper. Moreover, the applications of data mining in production companies are discussed, in particular the use of ANNs for the analysis of manufacturing data. Section 3 describes a case study of manufacturing process in a glass factory, defines the proposed approach to solving the problem under consideration, as well as shows the results of the approach. Section 4 provides information on the advantages and limitations of the approach.

2 Data Mining in Manufacturing

The constantly growing data resources are a challenge for manufacturing companies in every industry branch. This is due, inter alia, to a significant increase in the number of devices that generate data. Currently, the concept of Industry 4.0 and related technologies, such as Big Data or the Internet of Things (IoT), are very popular. These technologies facilitate the collection, processing, and use of large amounts of data. Data obtained from IoT can be exploited by many data mining techniques, including all data science methods known in the literature [1].

2.1 Knowledge Discovery from Data

Knowledge discovery from databases is a process whose task is a comprehensive data analysis, starting from a proper understanding of the problem under investigation, through the preparation of data, the building of appropriate models, and its evaluation [2]. Some variations of the knowledge discovery process are encountered in the literature. The most commonly used methodology among researchers is CRISP-DM. CRISP-DM defines the knowledge discovery process by dividing it into six phases [3]: (1) understanding the business domain from which the data comes; (2) detailed understanding of the data; (3) data preparation; (4) creating models; (5) evaluation of the obtained results; (6) implementation of the discovered knowledge in the business domain.

The first phase allows us to establish data mining goals in the context of the business domain. The second phase is the selection of appropriate data and the preliminary data analysis, which helps to understand the data [4]. Data sets obtained from industrial enterprises often contain incomplete, inconsistent, or erroneous data, which may be caused, for example, by failures in measuring or recording equipment. The third phase of CRISP-DM is to counteract these negative phenomena. In this phase, appropriate data preprocessing procedures are performed, such as removing observations containing erroneous data visible as outliers. Some methods of data analysis are sensitive to outliers; therefore, it may be important to remove or replace them [5]. Another problem encountered in the data is the missing observations resulting from improper work of the recording apparatus. If possible, missing observations should be replaced by using other observations from the data set. Otherwise, cases containing missing data must be omitted [6]. The stage of data pre-processing of consists also data transformation that results in converting the data to the appropriate types.

The fourth phase of CRISP-DM is modeling, the task of which is to search for certain dependencies, regularities, and patterns hidden in the data. The main tasks of data mining that appear most often in the literature are: detecting anomalies in data, modeling dependencies using association rules, object classification and clustering, regression, creating summaries in the form of reports and visualizations [2]. Generally, the methods used here can be divided into two groups: (1) classical statistical methods, e.g., linear regression, analysis of variance; (2) methods based on machine learning, such as classification and regression trees, ANNs, or support vector machines.

The fifth phase is the patterns interpretation of the obtained patterns and validation of the developed models. At this stage, it is checked whether the phenomena identified through exploration occur only in the analyzed data, or whether they can also be observed in a wider range of data. If the created models pass the validation phase, they can be deployed to the business domain and applied to new data (sixth phase).

2.2 ANNs and Production Data

For many years, the literature has witnessed a noticeable increase in the number of described applications of machine learning in production engineering. Such applications include, for example, the use of decision trees to determine the impact of production process parameters on the state of a blade [7], the application of random forests to investigate the influence of production process parameters on the number of defective products [8] as well as the application of decision trees in conjunction with various types of wavelets to predict the remaining useful life of cutting tools based on force and torque monitoring data [9]. Many papers have been published showing the wide range of possibilities offered by ANNs, among others. The work [10] shows the possibility of using data-driven models to predict failures of machines used in continuous technological processes. The aim of the research is to predict the total downtime of the machines involved in production.

A large part of the literature dealing with the use of data analysis in manufacturing companies is devoted to publications on product quality problems. The authors of the paper [11] point out that the prediction of the influence of the technological process parameters on the number of defects is very often based mainly on the experience and knowledge of the people working in the manufacturing company. In the problem of pressure casting, they propose models of ANNs that predict the influence of process parameters on the properties of products (e.g., clotting time, number of defects).

Article [12] deals with the production of galvanized steel. The aim of the research was to create a neural model that predicts the quality of steel expressed by its mechanical properties. The paper [13] describes the use of an ANN to model the relationship between turning parameters and the quality of the product expressed by the surface roughness. In turn, the authors of the paper [14] propose a model based on ANNs to predict the quality of products during turning on numerically controlled machine tools. In the article [15], the ANNs were used, inter alia, to identify key parameters that influence the quality of products manufactured in the injection molding process. The results of the research described in [16] show that approaches using ANNs give comparable results to other methods of quality control, including control charts and statistical techniques. The paper [17] refers to the glass coating process. It describes the use of ANN and the decision tree to model the relationship between these process parameters and the quality of the glass coating. The work [18] describes the use of a two-layer hierarchical ANN in the petrochemical industry. It was used to predict the quality of the product derived from the industrial pyrolysis of ethylene. The paper [19] describes the use of the bootstrap aggregation technique and multilayer perceptron ANNs to assess how the parameters of the manufacturing process affect the quality of the products.

In conclusion, the literature points out that in order to handle with the dynamic growth of data, there is a need to create tools that not only automatically collect data. These tools should be able to select the most appropriate data and apply proper analyzes to extract knowledge from the data [20]. More and more often, machine learning methods are replacing or supplementing classical methods. The main advantages of machine learning techniques are the ability to discover non-linear relationships in data and visualize interactions between variables. Moreover, machine learning techniques do not require meeting as many statistical assumptions in relation to the data as in the case of classical methods.

3 Research Problem

The research problem under consideration concerns the prediction of the number of defective products. The prediction was made based on data obtained from a glassworks that manufactures glass packaging. The data come from the production process monitoring system, which records the values of the selected parameters of this process. These are parameters related to:

- the operation of the blast furnace (e.g. glass level, glass temperature),
- cooling glass molds,
- atmospheric conditions in the production hall (atmospheric pressure, humidity, air temperature),
- the operation of the forehearth, which is divided into 6 sections (the data cover all sections and concerns, for example, the position of cooling valves or the temperature of the glass flowing through the forehearth).

In addition, data were obtained on the number of products with air bubbles in the glass, which is treated as a product defect. Data on the number of defective products and data on the values of the manufacturing process parameters come from the same period, which is 27 days. The values are recorded at intervals of 10 min.

3.1 Proposed Approach Based on Manufacturing Process in a Glassworks

The data obtained from the glassworks were divided as follows:

- The values of the production process parameters from production monitoring systems can potentially affect the number of defective products, therefore, they were treated as input (explanatory) variables,
- The number of defective products is a prediction target and therefore serves as an output variable (response variable).

The proposed approach to predicting the number of defective products is shown in Fig. 1. This approach is based on the CRISP-DM process; therefore, it begins with the business understanding phase. All the places from which the data serving as input variables were obtained are symbolically presented here. Each place is associated with an appropriate number of input variables (number of recorder parameters). The largest number of parameters obtained applies to the sixth section of the forehearth (15), and the smallest number of variables describes the atmospheric conditions and the cooling of the glass molds (3). In the first phase, contact with the company's employees is extremely valuable, because of them it is easier to understand the problem under study and to better select the process parameters that should be taken into account during the study.

The second phase was to understand the data and the relationship between the manufacturing process parameters. Understanding the data enables a better selection of data analysis techniques; therefore, it is also an important step in the proposed approach. In this phase, exploratory data analysis techniques, different types of visualizations (time series plots, histograms, box plots, quantile plots), and descriptive statistics were used. Research has shown that the acquired data are characterized by heterogeneous variance and the presence of several constant levels around which the values fluctuate. For some variables, outliers or missing observations were revealed. Furthermore, statistical tests and quantile plots did not show the consistency of the distribution of variables with the normal distribution. For this reason, classic methods of statistical data analysis should be used with caution, because some of them require the fulfillment of appropriate assumptions. Therefore, it was assumed that the posed problem of predicting the number of defective products would be solved by machine learning methods, which do not require meeting the assumptions about the normality of the distribution or homogeneity of the variance.

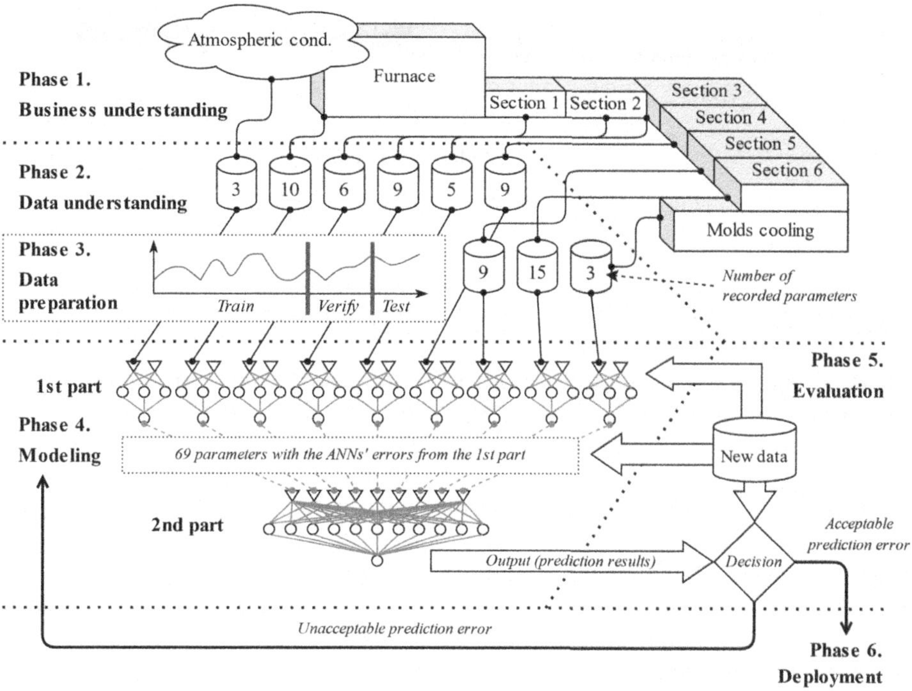

Fig. 1. Visualization of the proposed approach to the prediction of the number of defective products on the basis of production data.

The third phase consists of preparing the data for analysis (pre-processing). At this point, appropriate operations should be applied, for example, filling in missing observations or replacing outliers. The selection of these operations must depend on the data. These operations can be applied to both the input variables and the output variable.

Due to the fact that it was decided to use machine learning techniques, in the third phase, the data set should also be divided into three subsets: training, validation (verify), and test. Typically, machine learning methods assume a random division of the original data set into the three subsets, where each case is randomly assigned to one of the subsets. In the proposed approach, a different method of data division was used. In the problem under consideration, we deal with time series, so subsequent cases in the data set follow each other. Therefore, the prediction task will require predicting the next values based on the previous values. Random allocation of cases to the training, validation, and test sets could adversely affect the training process, because cases from the validation and test subset are not taken into account during training and would be treated as gaps in the training data. This would disrupt data continuity and the correct sequence of cases.

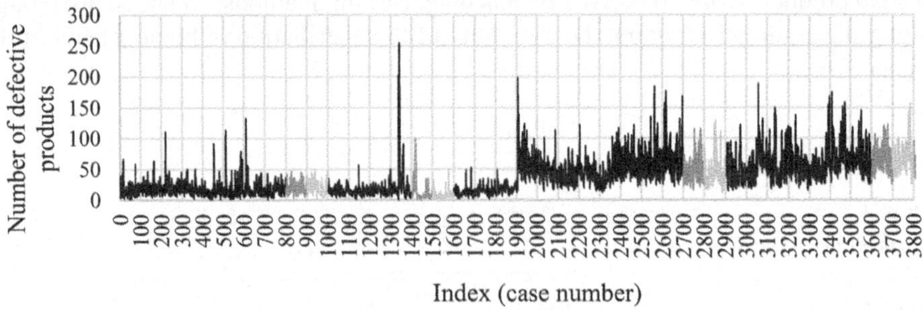

Fig. 2. Time series plot of the explained variable.

For the data under consideration, the following fragments of each variable were distinguished (see Fig. 2):

- Training subset: cases number 1–800; 1001–1400; 1601–2700; 2901–3600,
- Validation subset: 801–900; 1401–1500; 2701–2800; 3601–3700,
- Test subset: 901–1000; 1501–1600; 2801–2900; 3701–3800.

This division of the data was dictated by the course of the explained variable presented in Fig. 2. The first of the four training ranges covered the average level of the number of defective products. The validation and test ranges immediately following it are designed to test the predictive ability of the model to represent the average level of the dependent variable. The second training range covers slightly lower values of the explained variable, not counting single outliers.

The third training range includes a clear increase in the average level of the explained variable, and the validation and test ranges immediately following it are designed to check the predictive ability of the model to reflect the increased defectiveness of products. The highest average level of the dependent variable covers the fourth range of the training subset. The training subsets in total include about 80% of the cases, while the validation and test subsets contain approximately 10% of the cases each.

The fourth phase is the modeling stage. It was divided into two parts. The first part was about creating and training the nine ANNs. In the proposed approach, a separate ANN was created for each data source. Therefore, for each of the ANNs, only some of the input variables were used. The output variable for all ANNs was the number of defective products.

Due to the nature of the problem under consideration (prediction of time series), NARX ANNs were used. The visualization of the NARX network in an open form is shown in Fig. 3. During the training process, the NARX network takes into account the values of the explained variable $y(t)$ with the appropriate delay d, as well as the values of the explanatory variables $x(t)$ also with the delay d. For this reason, the predicted value of the explained variable y at time t is determined on the basis of d previous values of the variable y and d previous values of the explanatory variables x, which is expressed by formula (1):

$$y(t) = f(y(t-1), y(t-2), \ldots, y(t-d), x(t-1), x(t-2), \ldots, x(t-d)) \quad (1)$$

In the problem discussed, it was established, after consulting with the company's employees, that the number of defective products at a given moment can be most influenced by the values of the manufacturing process parameters from the period of one hour back. Given that the interval between consecutive observations is 10 min, it has been agreed that the delay d will be 6.

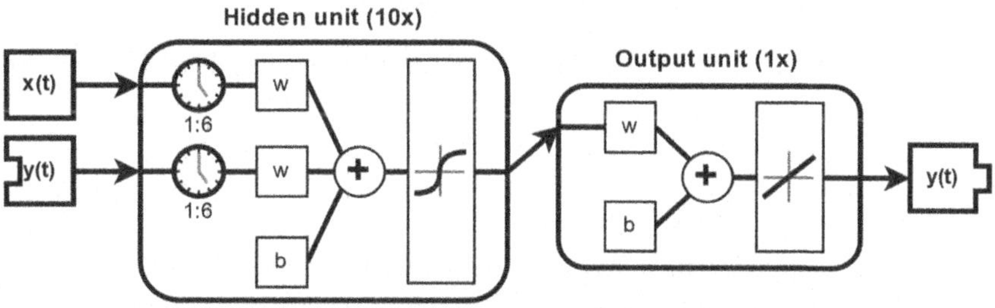

Fig. 3. NARX network in an open form, which was used during training on historical data.

The input values go to the neurons in the hidden layer by connections with appropriate weights. The hidden neuron first performs the operation of summing inputs multiplied by weights with bias. Then, the value of the activation function (in this case, the hyperbolic tangent) is calculated. For simplicity, it was assumed that each NARX network used has 10 hidden neurons. The calculated value of the activation function of hidden neurons goes to the output layer, where there is a neuron with the linear activation function. The task of the output neuron is to calculate the prediction result of the explained variable.

The training subset was used to train each NARX network. During training, the prediction error was calculated on the training, validation, and test subsets. The plot of the prediction error determined during the training process of one of the created NARX networks is shown in Fig. 4. The prediction error on the validation subset is used to prevent overfitting of the NARX networks. When this error increases its value in six successive epochs of the training algorithm, the training process of the NARX network is interrupted. The result of the learning process is the state of the NARX network at the moment when the validation error started to increase (the epoch denoted as Best in Fig. 4). In turn, the error calculated on the test subset was used in the second part of the modeling phase.

The role of the prediction error was played by mean squared error (MSE), which is one of the basic and most popular indicators for assessing models that generate numerical outputs. The MSE value is determined according to formula (2):

$$MSE = \frac{1}{n} \sum\nolimits_{i=1}^{n} (y_i - y_i^*)^2 \quad (2)$$

where n–number of cases in the data set, y_i–actual value of the explained variable for the i-th case, y_i^*–predicted value of the explained variable for the i-th case.

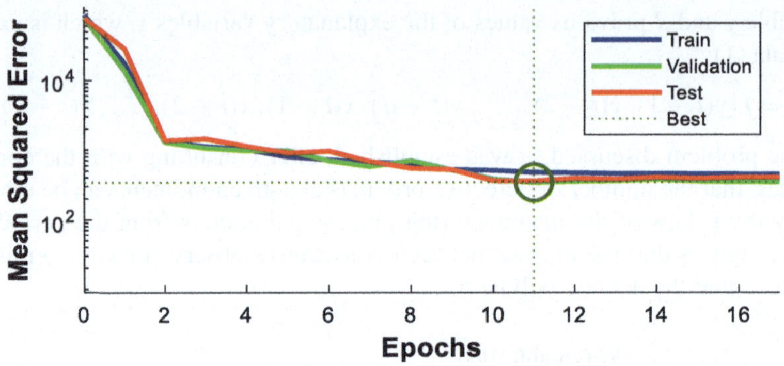

Fig. 4. The prediction error of one of the created NARX networks, calculated during the training process.

The second part of the modeling phase was to create and train a single NARX network. This time the NARX network took into account all 69 parameters of the manufacturing process, treating them as explanatory variables. However, the values of these variables were converted in a suitable manner before starting the training process. The transformation of the variables consisted of dividing each value of a given explanatory variable by the value of the prediction error calculated on the test subset for this NARX network from the first part of the modeling stage, which used the given variable as its input. In that way, the prediction errors computed in the first part were included in the second part of the modeling step. These errors act as weights that reduce the value of the explanatory variables. The higher the value of the prediction error of the NARX network from the first part of the modeling step, the more the values of the explanatory variables of this network were reduced. According to this approach, the explanatory variables, the use of which resulted in greater prediction errors in the first part, were reduced in the second part of the modeling step, so as to have less impact on the final prediction results of the number of defective products.

The fifth phase assesses the set of ANNs created in the fourth phase, which consists of ten NARX networks. For the evaluation of the set of ANNs, four data sets were used, which contain the values of all 69 explanatory variables and the explained variable. None of these data sets were used in the modeling stage.

The assumption of the fifth stage is to simulate the operation of the ANNs set in a situation in which a forecast of several future values of the number of defective products should be made. For this purpose, the structure of all NARX networks was changed to a closed form (see Fig. 5). The open NARX network allows for performing only one-step-ahead prediction. In turn, the closed form of the NARX network causes the network to contain a feedback loop. This gives the possibility to perform multi-step-ahead prediction. In this case, the prediction results of $y(t)$ will be used in place of the actual future values of $y(t)$.

The nine NARX networks established in the first part of the modeling phase were first evaluated. Each of these ANNs made predictions for the given values of the input variables, taking into account only the input variables that are assigned to the given ANN. The prediction of the number of defective products is made by each ANN six steps forward, taking into account the six previous values of the explanatory variables and the explained variable. The prediction error of each of the NARX networks is then calculated by comparing them with the actual values of the number of defective products. Prediction errors are treated as weights that reduce the values of the explanatory variables, similarly to the modeling phase: the values of the explanatory variables are divided by the prediction error of the corresponding NARX network.

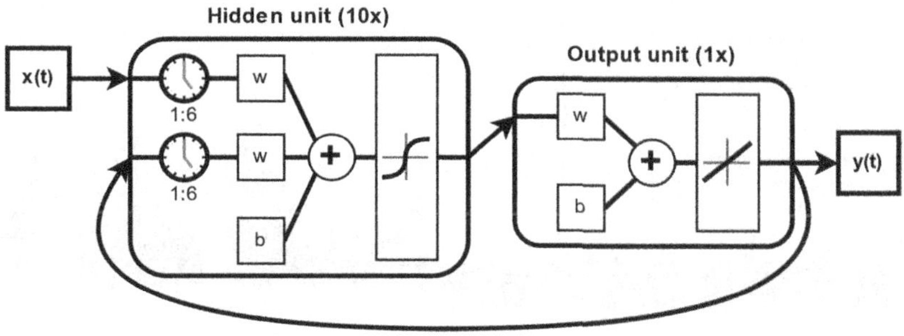

Fig. 5. NARX network in a closed form that was used in prediction for new data.

The data set prepared in this way goes to the final NARX network, which takes into account all explanatory variables. This network predicts the number of defective products six steps forward, taking into account six past values of the explanatory variables.

For the new data used in the evaluation phase, the actual number of defective products is known (although the NARX network does not take this into account when predicting). Therefore, it is possible to evaluate the predictions of individual NARX networks as well as the entire set of ANNs. If the prediction error is at an acceptable level, then the ANNs set can be deployed to operate in the business domain. However, if the prediction error is unacceptably high, then the created model should be modified and returned to the modeling phase. The modification may consist of selecting a different number of hidden neurons. If possible, it can also be extending the training dataset and re-training the ANNs.

3.2 Results and Conclusions

The quality of ANN depends not only on the training process and the training subset but also on the initial values of the weights of connections between neurons. The initial weight values are randomized and then modified in the ANN training process. The course of the training process can largely depend on the initial values drawn, and this, in turn, has a significant impact on the prediction ability of the ANN. Therefore, an approach to generate ten ANNs was used for each NARX network in the ANNs set. Each of the ten ANNs was trained using the same training subset. After completing the training, the quality of the ANNs was assessed using the test subset. The ANNs set included the ANN that had the smallest prediction error calculated on the test subset.

Figure 6 shows a comparison of the actual number of defective products with the number of defective products predicted by one of the NARX networks. The period included in Fig. 6 covers the entire training, validation, and test subset. It should be noted that the predictions in this case were generated by the NARX network in an open form. Therefore, the plot depicts the one-step-ahead predictions. Looking at the course of the time series of the dependent variable (upper part of Fig. 6) it can be seen that the considered NARX network makes the smallest errors in the case of the initial data fragment (up to approximately 1900 cases).

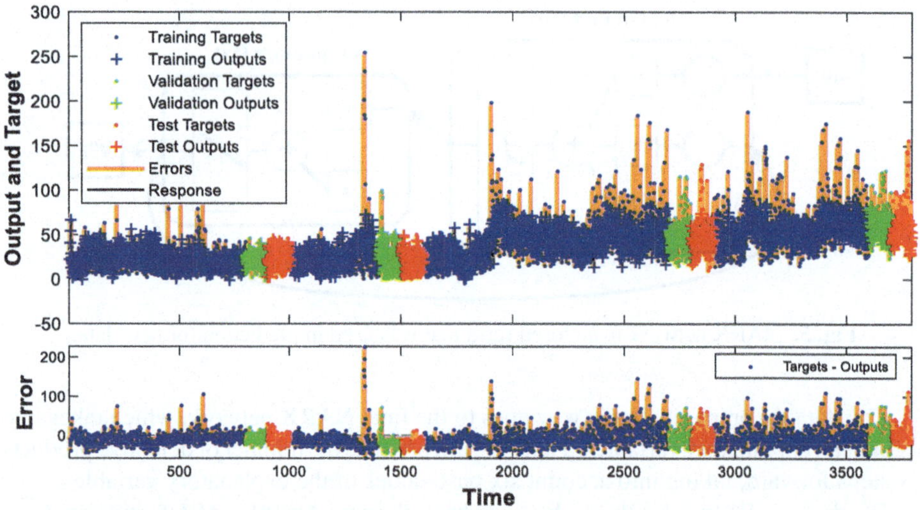

Fig. 6. Comparison of the actual values of the explained variable with the prediction results obtained by one of the NARX networks included in the ANNs set.

Later in the time series, the differences between the actual values and the predicted values are greater, which is clearly shown in the error graph at the bottom of Fig. 6. Moreover, it can be noticed that the NARX network in open form has a worse prediction of the highest values of the number of defective products. In such situations, the NARX network understates the prediction result. This observation is apparent for all three subsets: training, validation, and test. However, it is not the task of the neural model developed to predict outliers. Instead, the neural model created should estimate whether, with the given parameters of the manufacturing process, the number of defective products will remain at a low level or whether an increase in the number of defects will be observed. This task is accomplished by the NARX network because it is able to reproduce the increase in the number of defective products, which starts from approximately timestep number 2000. Moreover, it should be considered that the NARX network discussed comes from the first part of the modeling phase, which means that this ANN was trained only on a small subset of the manufacturing process parameters. Considering all the process parameters in the second part of the prediction modeling stage, the result of the prediction can be better.

One-step-ahead prediction, the results of which are shown in Fig. 6, usually does not reflect the real predictive ability of the created model, because in real applications the created model of the manufacturing process will be used to predict the number of defective products several timesteps ahead (e.g., for the next hour, which in this case means six-timesteps-ahead prediction). Therefore, a more reliable measure of the predictive ability of the model is the prediction error computed for the multi-step-ahead prediction. To determine the multi-step-ahead prediction, an additional test data set was used, which was divided into four periods. The prediction for these periods was performed by the NARX networks in closed form.

In addition, one additional NARX network (designation "single ANN") was created to serve as a benchmark. The single ANN was trained in an open form using the training subset, which included all the manufacturing process parameters. Thus, for the single ANN, all possible input variables were given without additional information about the significance of the input variables or their influence on the output variable. The single ANN in closed form was then used to predict four test periods. The single ANN prediction results were taken as the benchmark generated to compare the proposed approach with the single ANN approach. If the benchmark gives a prediction error close to the prediction error of the ANNs set or if the benchmark prediction error is smaller, then creating the ANNs set according to the proposed approach would be unjustified.

As with the rest of the NARX networks, also with the single ANN the approach was used, in which 10 ANNs were generated with the same parameters, and the ANN that made the smallest error on the test subset was selected. To make it easier to interpret the size of the prediction error, the RMSE (root mean squared error) measure was used, the result of which is expressed in the same units as the values of the explained variable. RMSE prediction errors calculated on the training, validation and test subsets of the best of the ten ANNs were 22.03, 18.61, 19.13, respectively. On the other hand, the corresponding measures for the NARX network from the second part of a modeling phase had the value of 18.09, 18.84, 18.87. Comparing the obtained RMSE errors, it can be concluded that the results for both approaches are similar. Therefore, there is no clear advantage of the proposed neural model over the NARX network treated as a benchmark. However, different conclusions can be drawn after analyzing the prediction results for the additional test data set, divided into four periods.

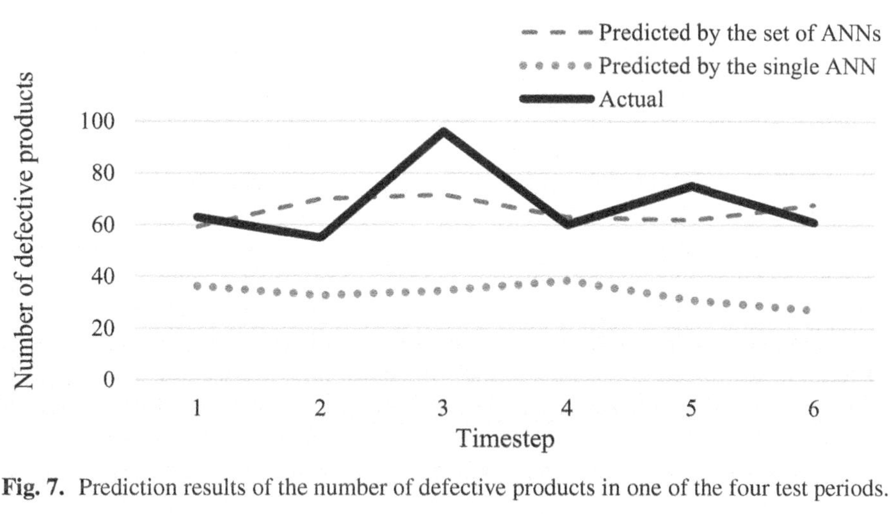

Fig. 7. Prediction results of the number of defective products in one of the four test periods.

Figure 7 shows the actual and predicted values of the explained variable in one of the four test periods. The forecast was made 6 steps forward. The results show a clearly better accuracy of the forecast made with the use of the ANNs set. Prediction by the single ANN underestimates the value of the explained variable at each of the six points. Also, in the remaining three test periods, the predictions made using the ANNs set are burdened with a smaller error compared to the benchmark single ANN. The RMSE prediction errors for all four test periods are shown in Table 1. In addition, Table 1 considers the prediction results of each of the nine NARX networks created in the first part of the modeling phase. For most of the rows, it can be seen that the RMSE values increase in subsequent periods. This may be due to the fact that each subsequent period is farther away from the training data.

Table 1. RMSE calculated for data from the four periods that were treated as test data.

ANN	1st period	2nd period	3rd period	4th period
Single ANN	30.10	37.87	36.69	42.37
Set of ANNs	16.14	13.32	15.28	19.98
Atmospheric cond.	16.70	21.86	23.62	27.59
Furnace	19.25	18.79	28.91	33.73
Section 1	14.91	17.92	27.19	30.94
Section 2	14.18	17.96	30.93	32.54
Section 3	15.52	21.92	28.99	32.52
Section 4	16.08	18.91	26.82	33.01
Section 5	15.03	18.99	33.04	35.15
Section 6	22.47	19.53	33.39	33.93
Molds cooling	25.47	22.15	29.66	34.09

Another interesting pattern can be seen in the case of the first period. The RMSE values generated for NARX networks based on forehearth sections number 1, 2, 3, 4, and 5 are lower than the RMSE of the ANNs set. This could suggest that the data from these five forehearths sections are sufficient to make the forecast. However, in the next three periods, the ANNs set has a clear advantage over NARX networks, with only the five sections listed. This example shows that when assessing the effectiveness of ANNs models, it is not worth relying on just one test period. The more possibilities to test the forecasts, the more reliable the evaluation of the created ANNs model.

The aforementioned research problem, which concerned the prediction of the number of defective products, is worth attention because by predicting this number, it is possible to determine how much glass cullet waste will appear after production. This allows the glassworks workers to better plan the demand for raw materials (glass cullet is one of the raw materials used). In addition, knowing the number of defective products, it is possible to more precisely determine the time needed to produce a given number of products.

Plans for further research may include more tests of the proposed approach, e.g. using data from other production lines in a glass factory or from other manufacturing companies. Furthermore, in the proposed approach, it is possible to add the step of selecting the appropriate number of hidden neurons in NARX networks, which could improve the prediction result. It also could be valuable to compare the predictive ability of the proposed ANNs model with other machine learning techniques (e.g., regression trees, random forests, support vector regression). In the proposed approach, for example, a random forest can be used instead of ANNs. Such an experiment could show whether the proposed approach is appropriate also for other types of models based on machine learning techniques.

4 Summary

The paper describes the process of creating a neural model of the production process on the example of a glassworks. The application of the created model to predict the number of defective products was presented on the basis of historical data on the values of selected manufacturing process parameters and the corresponding number of defective products. A characteristic feature of the proposed model is the creation of separate NARX ANNs for each stage of the manufacturing process. Each of the ANNs takes into consideration only that part of the manufacturing process parameters, which is assigned to a given stage of the process. The additional NARX ANN then considers all parameters along with the prediction results of previous ANNs.

The proposed neural model allows reducing the prediction error in comparison with the classic approach in which only one ANN is used, including all process parameters. Moreover, the proposed approach is independent of the business field. The presented example concerns the glass industry, but the approach can also be used in other industries or branches of industry. It seems that for each research problem in which the relationship between the input variables divided into certain subgroups and the quantitative output variable is sought, the proposed approach can be applied.

The time devoted to building the neural model may be a limitation of the proposed approach. The neural model contains many separate ANNs, each of which must first be trained and then tested. This takes more time than the classic approach based on just one ANN. The second limitation is the number of historical values that must be available to make a forecast. In the example of a glassworks considered, it was assumed that the six previous values of the manufacturing process parameters and the number of defective products should be considered, because the previous six values have the greatest impact on the future number of defective products. Due to the fact that the proposed model works in two stages, it is necessary to have twelve previous values of the parameters and the number of defective products. In the case of the classic approach with one ANN, it is sufficient to have only six previous values of the parameters mentioned.

References

1. Taşıran, A.C.: Internet of things and statistical analysis. In: Al-Turjman, F. (eds.) Performability in Internet of Things. EAI/Springer Innovations in Communication and Computing, pp. 127–136. Springer, Cham (2019). https://doi.org/10.1007/978-3-319-93557-7_8

2. Larose, D.T., Larose, C.D.: Discovering Knowledge in Data: An Introduction to Data Mining. Wiley, Hoboken, New Jersey (2014)
3. Chapman, P., et al.: CRISP-DM 1.0: step-by-step data mining guide. Computer Science (2000)
4. Witten, I.H., Frank, E., Hall, M.A., Pal, C.J.: Data mining: Practical Machine Learning Tools and Techniques, 4th edn. Morgan Kaufmann Publishers Inc., San Francisco (2016)
5. Čížek, P., Sadıkoglu, S.: Robust nonparametric regression: a review. WIREs Comput. Stat. **12**(3), 1–16 (2020)
6. Amiri, M., Jensen, R.: Missing data imputation using fuzzy-rough methods. Neurocomputing **205**, 152–164 (2016)
7. Antosz, K., Mazurkiewicz, D., Kozłowski, E., Sęp, J., Żabiński, T. Machining process time series data analysis with a decision support tool. In: Machado, J., Soares, F., Trojanowska, J., Ottaviano, E. (eds.) Innovations in Mechanical Engineering. ICIENG 2021. LNME. Springer, Cham (2022). https://doi.org/10.1007/978-3-030-79165-0_2
8. Setlak, G., Pasko, L.: Random forests in a glassworks: knowledge discovery from industrial data. In: Świątek, J., Borzemski, L., Wilimowska, Z. (eds.) ISAT 2019. AISC, vol. 1051, pp. 179–188. Springer, Cham (2020). https://doi.org/10.1007/978-3-030-30604-5_16
9. Kozłowski, E., Antosz, K., Mazurkiewicz, D., Sęp, J., Żabiński, T.: Integrating advanced measurement and signal processing for reliability decision-making. Eksploatacja i Niezawodnosc Maint. Reliab. **23**(4), 777–787 (2021)
10. Merh, N.: Applying predictive analytics in a continuous process industry. In: Laha, A. (eds.) Advances in Analytics and Applications. Springer Proceedings in Business and Economics. Springer, Singapore (2019). https://doi.org/10.1007/978-981-13-1208-3_10
11. Krimpenis, A., Benardos, P.G., Vosniakos, G.-C., Koukouvitaki, A.: Simulation-based selection of optimum pressure die-casting process parameters using neural nets and genetic algorithms. Int. J. Adv. Manuf. Technol. **27**(5), 509–517 (2006)
12. Meré, J.B.O., Marcos, A.G., González, J.A., Rubio, V.L.: Estimation of mechanical properties of steel strip in hot dip galvanising lines. Ironmak. Steelmak. **31**(1), 43–50 (2004)
13. Lin, W.S., Wang, K.S.: Modelling and optimization of turning processes for slender parts. Int. J. Prod. Res. **38**(3), 587–606 (2000)
14. Suneel, T.S., Pande, S.S., Date, P.P.: A technical note on integrated product quality model using artificial neural networks. J. Mater. Process. Technol. **121**(1), 77–86 (2002)
15. Ali, I.G., Chen, Y.T.: Design quality and robustness with neural networks. IEEE Trans. Neural Netw. **10**(6), 1518–1527 (1999)
16. Ozcelik, B., Erzurumlu, T.: Comparison of the warpage optimization in the plastic injection molding using ANOVA, neural network model and genetic algorithm. J. Mater. Process. Technol. **171**(3), 437–445 (2006)
17. Li, M., Feng, S., Sethi, I.K., Luciow, J., Wagner, K.: Mining production data with neural network & CART. In: Third IEEE International Conference on Data Mining, pp. 731–734 (2003)
18. Zhou, Q., Xiong, Z., Zhang, J., Xu, Y.: Hierarchical neural network based product quality prediction of industrial ethylene pyrolysis process. In: Wang, J., Yi, Z., Zurada, J.M., Lu, B.L., Yin, H. (eds.) Advances in Neural Networks - ISNN 2006. ISNN 2006. LNCS, vol. 3973, pp. 1132–1137. Springer, Berlin, Heidelberg (2006). https://doi.org/10.1007/11760191_165
19. Paśko, Ł, Kuś, A.: Bootstrap aggregation technique for evaluating the significance of manufacturing process parameters in the glass industry. Tech. Sci. **24**, 135–155 (2021)
20. Peters, H., Link, N.: Cause & effect analysis of quality deficiencies at steel production using automatic data mining technologies. IFAC Proc. Vol. **43**(9), 56–61 (2010)

The Role and Meaning of the Digital Twin Technology in the Process of Implementing Intelligent Collaborative Robots

Jakub Pizoń[1], Arkadiusz Gola[2](✉), and Antoni Świć[2]

[1] Faculty of Management, Department of Enterprise Organization, Lublin University of Technology, Lublin, Poland

[2] Faculty of Mechanical Engineering, Department of Production Computerisation and Robotisation, Lublin University of Technology, Lublin, Poland

a.gola@pollub.pl

Abstract. In the last decade, the paradigms concerning designing and enhancing manufacturing systems were changed due to an uncommonly fast development of computerization, automation and robotization. Unknown information technologies, such as virtual reality, artificial intelligence-based solutions and robots enter the industrial reality to an increasingly great extent. Development trend analysis allows one to assume that in the nearest future collaborative robots, capable of intelligent cooperation with people will constitute an element of manufacturing systems. The objective of this article is to present conditions that ought to be met in order to introduce intelligent collaborative robots as well as to employ Digital Twin technology in the process of cobot adaptation in terms of a certain manufacturing system.

Keywords: Manufacturing systems · Digital twin · Cobots · Artificial intelligence · Virtual reality

1 Introduction

Industry 4.0 revolutionised the production sector by integrating and implementing information technologies such as artificial intelligence (AI) Internet of Things (IoT), cloud computing or cyber-physical systems (CPS) [1, 2]. The main rule of Industry 4.0 is to make the manufacturing system "intelligent" by connecting machines and devices, which can control each other throughout their entire life cycle [3, 4]. The main priority of Industry 4.0 is process automation and therefore limiting human contribution to the manufacturing process [5, 6]. Industry 4.0 concentrates on enhancing mass production and effectiveness by providing intelligent connection between devices and applications using machine learning [7, 8]. Moreover, the fourth industrial revolution brought companies' attention to the fact that data-driven management offers certain possibilities [9]. Data aggregated in real time present changing of production in overtime [10].

B. Gapiński et al. (Eds.): MANUFACTURING 2022, LNME, pp. 39–49, 2022.
https://doi.org/10.1007/978-3-031-00805-4_4

The more possibilities of measuring and analysing, the greater potential for optimizing. It is therefore to be stated that data aggregation in itself does not determine competitive advantage, but is a step towards a possibility of a full usage of production data potential [11]. It is also worth mentioning that each company is different and may collect a different range and volume of data [12]. The properly analysed data become a basis for management decision-making in manufacturing [13]. Due to this fact numerous question occur in terms of data analysis, mainly on selecting proper algorithms and determining the method of data aggregation and measure definition, that would allow one to assess the occurring changes [14]. Answering those questions poses a significant difficulty. Prior to answering, it is necessary to obtain a base – a place for data implementation – to obtain a coherent result.

The objective of this article is to present conditions and strategies resulting from current development trends in industrial systems, which influence and are used for implementing cobots. The possibility of using Digital Twin technology in the adaptation of cobots for certain manufacturing systems was especially examined.

2 Digital Twin as an Environment for Designing and Improving Modern Manufacturing Systems

2.1 General Idea of Digital Twin Technology

As far as optimization of processes and manufacturing systems, a concept of Digital Twin, promoted in terms of Industry 4.0 may prove helpful. Digital Twin is a structured method of digitalizing manufacturing systems. Digital Twins are defined as a digital representation of an active, unique product, service or manufacturing system characterized by certain characteristics or conditions used for analysing, understanding and improving the product, service or manufacturing system [15]. According to the research by Markets and Markets, the worth of the Digital Twins global market of Digital Twins was stated to be 3.1 billion dollars in 2020, whereas in 2026 it is estimated to reach 48.2 billion dollars [16].

Digital Twin can be characterized by three basic elements: physical reality, virtual representation and interconnections that exchange information between the physical reality and virtual representation. It was presented in Fig. 1 in schematic form.

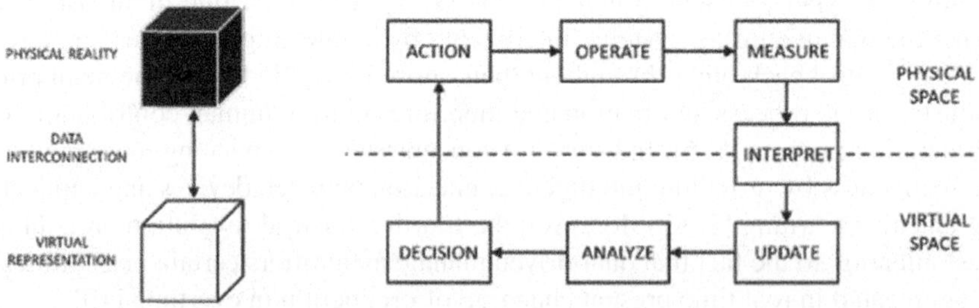

Fig. 1. Digital Twin components and high-level processes [17].

The main motivation for the development of solutions on the level of Digital Twin is to improve the safety of manufacturing systems [18]. In this case it is focused on automatic fault-tolerant control, FTC. This structure allows process manufactures to continue predefined work in the case of equipment failure. The work may not be as efficient as normal, but the manufacturing system does not cease to operate [19].

2.2 Benefits of Digital Twin Implementation

The solutions of the Digital Twin are useful when a diagnose as well as prognosticating failures/errors of physical devices, planning movement by integrating with agent-based algorithms of fast optimization in a robotized production hall, modelling the real state of objects in order to increase their reliability, productive capacity or to plan the path of the robotic tool on the basis of transferring information between a reconfigured additive manufacturing system and its digital twin [16].

Management feedback using a digital twin is conducted by creating a model and its verification by what-if testing with another model proposition/suggestion. In the feedback phase of the digital twin, virtual start with the what-if tests combines simulation and automation. Virtual production modelling offers numerous benefits presented below [20].

1. Finite production planning and task schedule testing to avoid unplanned production downtime.
2. Supporting decision-making ‘to optimize production.
3. Avoiding human errors in real production by virtual simulation and testing.
4. Allowing for easy error/problem detection in the manufacturing system.

Environment and context area of a Digital Twin is represented by four dimensions: integration breadth, connection mode, update frequency and product life cycle.

Digital Twin behaviour or the wealth of opportunity include the remaining four dimensions, mainly CPS intelligence, simulation capabilities, digital model richness and human interaction. An eight-dimension model determined in such a way can help in the process of setting goals in the development of a new product by using the existing DT, upgrading the already existing product with the knowledge obtained from their operative DT, using standardization tool for the further development of DT as a single product or service (as a template), with the possibility of adding new functions along the eight dimensions as needed [15].

3 The Issue of Implementing Cobots in the Light of Current Trends

3.1 Motivation for Implementing Intelligent Collaborative Robots

Researchers indicate that the fourth industrial revolution decreases the need for a human component/factor in production, as it focuses on automation and seeking algorithm-based optimization. It is also frequently stated that numerous social consequences may ensue and the possibility of innovation is decreased, since it is an element characteristic for humans, as opposed to machines. Replacing humans with machines became naturally an area of research on new ways of a mutual (safe and effective) cooperation between human and machine. The term "cobot", describing a robot able to work and cooperate with humans, has been introduced.

Cobot was designed (both in terms of construction and application/usage) to be safe operating in the vicinity of humans [21]. Moreover, the term "intelligent collaborative robot" can be found in literature more,21,and more frequently. Cobots are equipped with elements of artificial intelligence allowing one to broaden the range of intelligent interaction with objects nearby [22–24].

Introducing a human factor in the form of intelligence to the realization of processes with cobots is stated to be one of the elements of the fifth industrial revolution (Industry 5.0). The essence of Industry 5.0 is currently in the phase of concept formulation and is focused on the possibilities of using human intelligence and creativity to cooperate with powerful, intelligent and precise machines.

Numerous technical visionaries believe that Industry 5.0 will allow not only for using intelligence, but also the sense of touch (characteristic for humans) in the manufacturing industry. Industry 5.0 is expected to combine the high speed and precision of machines with critical and cognitive human thinking. Mass personalization is another important contribution of Industry 5.0, in which the customers may prefer personalized products, which suit their needs and taste. Industry 5.0 shall significantly increase production effectiveness and create versatility between humans and machines, allowing for responsibility in terms of interaction and constant monitoring [7].

Industry 5.0 acknowledges the industry's ability to reach social goals exceeding workplaces and growth so as to support sustainable development (respecting the capability of the environment while concentrating on the wellbeing of the industry worker) [25]. This goal comprises of these three basic elements:

- human-centric approach – instead of treating new technologies as a starting point and researching its potential for increasing effectiveness, human-centric approach places human needs and interests in the centre of the production process. Instead of focusing on the possible applications of the new technology, we ask what it can do for people;
- sustainability; sustainability means decreasing energy consumption and greenhouse gas emission in order to avoid depletion and degradation of natural resources to fulfil the needs of current generations without jeopardizing the future generations;
- resilience – focuses on the need to develop a higher level of resilience in manufacturing, bracing it against disruptions and ensuring its readiness to support critical infrastructure in the times of crisis [25].

A new perspective of developing manufacturing systems is introduced considering these three elements. This perspective would bring people back to factory halls, connecting people and machines to utilize the power and creativity of human mind in order to increase process effectiveness by combining the workflow with intelligent systems [26].

3.2 Cobots – Collaborative Robots

Industry 5.0 redefines the notion "robot". Robots shall no longer be simply programmable machines suited for performing repetitive actions. Instead they could become a perfect companion for people in certain cases. The next industrial revolution will introduce a next generation of robots – cobots, which will know or learn quickly how to perform tasks.

These collaborative robots will be aware of human presence and therefore shall take care of safety and risk criteria. They will be able to notice, understand and feel not only a person, but also the goals and expectations of a human operator. Similarly to a student, cobots will observe and learn how to perform a certain task. Once the task is learned, it will be performed by the cobots as it was done by humans [25].

3.3 Cobots – Collaborative Robots

Despite a promising perspective presented by Industry 5.0, there is a lot to be done in terms of robot implementation, and especially its conditions. According to the authors, numerous requirements presented in literature ought to be met by Industry 5.0. These conditions can be divided into two main groups: technical and technological conditions and organisational conditions.

The first group is focused on technical and technological measures that ought to be used in creating solutions for Industry 5.0 and cobots. It is necessary for the solution to be cohesive in terms of the working environment for both humans and machines. It is also indicated that in order to consider such a solution it is necessary to present ready methods and tools in terms of both obtaining and processing data. Technologies supporting the implementation of cobots is a set of complex systems which combine technologies, such as intelligent materials with biology-inspired sensors. Therefore, each of the below presented elements can reach their potential only when combined with others, as a part of a system or a technological solution.

- *networked sensor data interoperability* – ensuring the possibility of data flow between various machines and sensors (as required), energy-saving and safe technologies of transmission, storing and analysing of data,
- *multiscale dynamic modelling and simulation: digital twins* – allowing one to map the production process in the virtual reality, digital twns and simulation technologies optimize production, test products and processes as well as detect possible consequencs,
- *shopfloor trackers* – constituting a base for robot navigation in the production area;
- *virtual training* – training in a simulated environment, allowing one to select a proper solution for the type of work, staff education,
- *intelligent autonomous systems* – working without an operator, deep learning necessary for pattern recognition, ability to deal with and find correlations between complex, interconnected data from various sources and scales in terms of a system of systems,
- *advances in sensing technologies and machine cognition* – a base for recognizing patterns and applying them in machines,

- *applying brain monitoring technology in order to recognize signals* – encephalography (EEG). Functional Magnetic Resonane Imaging (fMRI) or functional near-infrared spectroscopy (fNIRS) (fNIRS is portable and easier to use due to a shorter configuration time and built-in wireless connection for transferring data in the majority of available headsets [26],
- *individualised human-machine-interaction* –solutions allowing one to combine the work done by humans with technology, support people and combining human innovation with the capability of a machine, recognizing speech and gestures, predict intentions, control mental and physical burden and stress in workers, cobots, augmented, virtual or mixed reality, exoskeletons, bio-inspired work clothes and safety equipment,
- *bio-inspired technologies and smart materials* – bio-inspired technologies and processes ensured by built-in sensors, adaptative-ergonomic properties of the surface,
- *technologies for energy efficiency, renewables, storage and autonomy* – technologies require significant amounts of energy to properly function, therefore solutions obtaining carbon neutrality are required, mainly integration of renewable energy sources, Hydrogen and Power-to-X technologies support, intelligent dust sensors, energetically autonomous or low-energy data transmission and analysis [27].

The other group of conditions are organizational issues. Here, the issues regarding the influence of cobots on the work environment, employees, organizational structure and ethical or legal issues are discussed. The following issues are presented:

- *legal and regulatory issues:*
  There is still no coherent legal definition of a robot. Technical definitions are available, but only a legal definition is a basis for business and organizations. Moreover, the definition ought to allow one to discern a robot from other machines occurrin in the production environment. Apart from a clear definition, the law ought to consider at least the types of robots, which might be operated alongside human workers, roles and responsibilities of robots, types of decisions concerning humans that can be made by robots, types of malfunctions and responsibilities, distribution of responsibility in the even of robot malfunction, robot software regulations, the robot's ability to learn adn develop, certificates necessary, certificating bodies and their responsibilities;
- *personal preference toward working with robots:*
  Some workers may be interested in working with robots, whereas some others might be against it. Organizations interested in implementing such solutions ought to be aware of those preferences, since they will influence the speed of implementation and the number of problems relating to it;
- *psychological issues resulting from human-robot co-working:*
  It is difficult to determine how human psyche will be influenced by the cooperation with robots. Could it cause phobias? At this point it can be stated that further research will be necessary;
- *social implications of human-robot co-working:*
  During work, numerous social interactions occur, also in a group. Some people might prefer robots presenting social behaviour. Others might claim that social behaviour

exhibited by robots is unreal and, since it results from computer programming, false. The employees' opinions will most likely vary;

- *the changing role of human resources departments:*
  Conducting work analyses, creating descriptions of workstations and identification of business processes are the most important tasks of the human resources department. HR departments will have to conduct their tasks again before introducing robots to their organizations;
- *the changing role of information technology departments and emerging of robotics departments:*
  Since robots take over important tasks, their conservation and modernisation ought to be performed in a safe and secure manner. It is important to ensure security of data obtained by the robots. This shall pose even more difficulty than securing the current IT infrastructures of an organization. Cooperating with robots will make it harder to ensure information and privacy in the workplace;
- *ethical issues resulting from human-robot co-working – ethical status of robots:*
  Commitment to work and a proper performance of duties are directly linked to humans and their work. In the case of robots, this ethics does not apply, since they perform certain strictly defined tasks. For this reason, people cannot compete with robots in terms of obtaining a good ethical norm in this aspect. It is hard to predict how work ethics will evolve in the area of human-robot cooperation;
- *preference toward types of robots to work with (learning or rule-based robots):*
  Robot learning method depends on the applied technique. It can either synthesize knowledge in terms of machine learning, or work according to strict rules. As a result, we will be faced with a compromise in terms of the types of robots used in the environments where humans cooperate with robots;
- *learning to work with robots:*
  We may be forced to learn how to behave towards a robot in order to achieve what we want. Non-verbal communication is an important part of interpersonal interactions. Robots can, but do not necessarily have to understand these nonverbal signals. People will take time to learn and get used to working with robots;
- *negative attitude toward robots due to shrinking human workforce:*
  Some might claim that using robots may increase unemployment. A counterargument, however, indicates that additional workplaces will be created. History validates this counterargument. Since machines would take over more "earthbound" tasks, people could focus on jobs requiring creativity, artistry, research and development;
- *humans competing with robots or robots complementing humans:*
  The majority of people may even prefer obedient robot-assistants. Such robots would help people with their tasks without a need for a human assistant. On the other hand, some people may claim that that robots can, and even should take up jobs for people, as it would increase the effectiveness of human workers. People compete for jobs in organizations and adding robots to this competition would complicate management and organisation [28].

A production system changes significantly as cobots are introduced to it. The awareness of organizational conditions of robots must be taken into account and might help to assess whether a solution would be beneficial for a certain production system.

3.4 The Role of Digital Twin in the Implementation of Cobots for Production Systems of Industry 5.0

Among numerous conditions relating to implementing cobots an area requiring attention occurs, mainly implementing the robot in certain industrial conditions. It is important to realise that the human-robot cooperation poses a challenge for both the cobot builders and the people operating the machines due to the variability of people. Even in the case of an identical task, each movement of a person performing it might be slightly different. On the other hand, cobots might perform various tasks and ae introduced to various production contexts. The cobot activity depends on the data availability and the quality of controlling algorithms. The more data, the better possibilities for robot parametrization, which means a better human-robot cooperation. Work safety increases similarly in the case of anormal or atypical situations.

In the case of purchasing a new cobot equipped with specialist sensory infrastructure (cameras, laser sensors etc.) and pre-defined algorithms it is necessary to adjust the cobot to the needs of a given organization.

Two strategies appear to be especially promising. In the first one, the cobot is implemented directly into the work environment, which would allow the algorithms to parametrize the behaviour on the basis of the sensors available. In this case, however, the cobot will be limited to the data it can obtain itself. In many cases this would suffice, but in the case of a complex machinery park or a production line that is constantly in motion (the workers moving, automated guided vehicles supplying materials etc.) it is impossible for a cobot to notice everything using cameras and sensors. For example, cobot will not be able to notice a person, whose view is obstructed by another person or an object.

For this reason and considering the development of Digital Twin solutions it is advised to consider the second, much more complex strategy, where the cobot is implemented using a Digital Twin. Moreover, the environment is characterised using the data from Digital Twins. This way, almost the entirety of the environment is represented by digital data – both the human workers and the machines have a representation in the virtual reality (Fig. 2). Data on the full spectrum of events in the workplace allows one to implement a greater number of boundary conditions. Depending on the needs, fulfilling the conditions can be seen as a safety or work effectiveness operations. In the case of humans, human robot collaboration approach is implemented and the human digital twin solutions are used [21].

In the case of machines, Digital Twin solutions are usually used. This approach offers a greater range of data, which allows both the robot builder and the robot itself to perform tasks both at the implementation stage and in everyday work in industrial conditions.

In the implementation stage (even before the physical implementation) it is possible to train the cobot on the basis of the aggregated real production data. The training is therefore not based on simulated or estimated data, but on data registered in the place where the robot is expected to operate. Such an approach shortens the implementation time and limits the risk of human-machine cooperation.

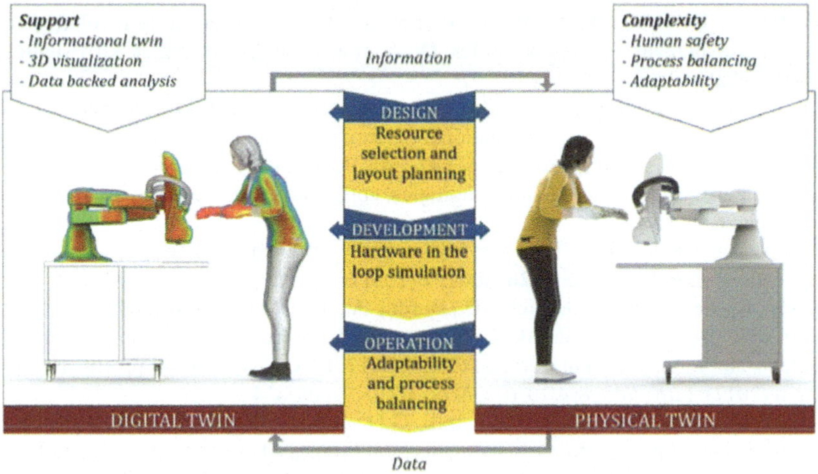

Fig. 2. Information flow between digital and physical human-twin [21].

Upon completing the implementation period, new production launches are performed elastically. Cobot ought to be able to learn. In the case when data is supplied by Digital Twins, such an adaptation is based on the analysis of new data, which is in accordance with the current production requirements. In this, way the cobot becomes a fully functional tool, an open system.

4 Conclusions

Considering the above discussed conditions and strategies of implementing cobots, it is to be mentioned that this issue is of great importance for the development of production systems. It is necessary to determine how a certain solution realises technical, technological and organizational conditions when developing, planning or implementing such a solution. It is a basic step towards creating a fully functional solution meeting the needs of the robot builders, companies and production workers. The role and meaning of Digital Twin technology and cobots shall be researched further.

Acknowledgments. The project/research was financed in the framework of the project Lublin University of Technology-Regional Excellence Initiative, funded by the Polish Ministry of Science and Higher Education (contract no. 030/RID/2018/19).

References

1. Gola, A.: Design and management of manfufacturing systems. Appl. Sci. **11**, 2216 (2021)
2. Grznar, P., et al.: Modeling and simulation of processes in a factory of the future. Appl. Sci. **10**(13), 4503 (2020)
3. Jasiulewicz-Kaczmarek, M., Żywica, P., Gola, A.: Fuzzy set theory driven maintenance sustainability performance model: a multiple criteria approach. J. Intell. Manuf. **32**(5), 1497–1515 (2021)

4. Petrillo, A., De Felice, F., Cioffi, R., Zomparelli, F.: Fourth industrial revolution: current practices, challenges, and opportunities. Digit. Transform. Smart Manuf. (2018).https://doi.org/10.5772/intechopen.72304
5. Kluz, R., Antosz, K.: Simulation of flexible manufacturing systems as an element of education towards Industry 4.0. In: 6th International Scientific-Technical Conference on Advances in Manufacturing II (Manufacturing), pp. 332–341. Poznan (2019)
6. Lu, Y.: Industry 4.0: a survey on technologies, applications and open research issues. J. Ind. Inf. Integr. **6**, 1–10 (2017)
7. Maddikunta, P.K.R., et al.: Industry 5.0: a survey on enabling technologies and potential applications. J. Ind. Inf. Integr. **26**, 100257 (2021)
8. Musial, K., Kotowska, J., Górnicka, D., Burduk A.: Tabu search and greedy algorithm adaptation to logistic Task. In: Saeed, K., Homenda, W., Chaki, R. (eds.) Computer Information Systems and Industrial Management. CISIM 2017. LNCS, vol. 10244, pp. 39–49. Springer, Cham (2017). https://doi.org/10.1007/978-3-319-59105-6_4
9. Arrais-Castro, A., Varela, M.L.R., Putnik, G.D., Ribeiro, R.A., Machado, J., Ferreira, L.: Collaborative framework for virtual organisation synthesis based on a dynamic multi-criteria decision model. Int. J. Comput. Integr. Manuf. **31**(9), 857–868 (2018)
10. Brzozowska, J., Gola, A.: Computer aided assembly planning using MS excel software – a case study. Appl. Comput. Sci. **17**(2), 70–89 (2021)
11. Rymarczyk, T., Klosowski, G., Kania, K., Rymarczyk, P., Mazurek, M.: Tomographic ultrasonic sensors in insutrial applications. Przeglad Elektrotechniczny **97**(1), 166–169 (2021)
12. Kłosowski, G., Gola, A., Thibbotuwawa, A.: Computational intelligence in control of AGV multimodal systems. IFAC-Papers Online **51**(11), 1421–1427 (2018)
13. Relich, M., Nielsen, I.: Estimating production and warranty cost at the early stage of a new development project. IFAC Papers OnLine **54**(1), 1092–1097 (2021)
14. Pizon, J., Kulisz, M., Lipski, J.: Matrix profile implementation perspective in industrial internet of things production maintenance application. J. Phys. Conf. Ser. **1736**(1), 01236 (2021)
15. Stark, R., Fresemann, C., Lindow, K.: Development and operation of Digital Twins for technical systems and services. CIRP Ann. **68**(1), 129–132 (2019)
16. Leng, J., Wang, D., Shen, W., Li, X., Liu, Q., Chen, X.: Digital twins-based smart manufacturing system design in industry 4.0: a review. J. Manuf. Syst. **60**, 119–137 (2021)
17. VanDerHorn, E., Mahadevan, S.: Digital twin: generalization, characterization and implementation. Decis. Support Syst. **145**, 113524 (2021)
18. Stączek, P., Pizoń, J., Danilczuk, W., Gola, A.: A digital twin approach for the improvement of and autonomous mobile robots (AMR's) operating enviroment – a case study. Sensors **23**(21), 7830 (2021)
19. He, R., Chen, G., Dong, C., Sun, S., Shen, X.: Data-driven digital twin technology for optimized control in process systems. ISA Trans. **95**, 221–234 (2019)
20. Jeon, S.M., Schuesslbauer, S.: Digital twin application for production optimization. In: IEEE International Conference on Industrial Engineering and Engineering Management (IEEM), pp. 542–545. IEEE, Singapore (2020)
21. Malik, A.A., Brem, A.: Digital twins for collaborative robots: a case study in human-robot interaction. Robot. Comput. Integr. Manuf. **68**, 102092 (2021)
22. Doyle-Kent, M., Kopacek, P.: Industry 5.0: is the manufacturing industry on the cusp of a new revolution? In: Durakbasa, N., Gençyılmaz, M. (eds.) Proceedings of the International Symposium for Production Research 2019. ISPR (ISPR 2019) 2019. LNME, pp. 432–441. Springer, Cham (2020). https://doi.org/10.1007/978-3-030-31343-2_38
23. Castillo, J.F., Ortiz, J.H., Diaz Velasquez, M.F., Saavedra, D.F.: COBOTS in Industry 4.0: safe and efficient interaction, collaborative and humanoid robots. https://www.intechopen.com/online-first/77896. Accessed 06 Dec 2021

24. Saxena Dev, S.V.A., Saxena, A.: Emergence of futuristic HRM in perspective of human-cobot's collaborative functionality. Int. J. Eng. Adv. Technol. **10**, 2249–8958 (2021)
25. Javaid, M., Haleem, A.: Critical components of industry 5.0 towards a succesful adoption in field of manufacturing. J. Ind. Integr. Manag. Innov. Enterp. **5**(3), 327–348 (2020)
26. Nahavandi, S.: Industry 5.0 - a human-centric solution. Sustainability **11**(16), 4371 (2019)
27. Müller, J.: Enabling Technologies for Industry 5.0 - Results of a workshop with Europe's technology leaders (2020). https://op.europa.eu/en/publication-detail/-/publication. Accessed 06 Dec 2021
28. Demir, K.A., Döven, G., Sezen, B.: Industry 5.0 and human-robot co-working. Procedia Comput. Sci. **158**, 688–695 (2019)

Workspace Analysis of a Novel Parallel Kinematic Machine with 6 Degrees of Freedom

Florin Popişter[1](✉), Alexandru Oarcea[2], Sergiu-Dan Stan[2], and Costan-Vlăduţ Trifan[1]

[1] Faculty of Industrial Engineering, Robotics and Production Management, Technical University of Cluj-Napoca, Cluj-Napoca, Romania
florin.popister@muri.utcluj.ro

[2] Faculty of Automotive, Mechatronics and Mechanical Engineering, Technical University of Cluj-Napoca, Cluj-Napoca, Romania

Abstract. This paper presents research within are studied and applied distinct types of determination of the workspace of a 6 DOF Parallel Kinematic Machine (PKM). PKM structures assume the incorporate of flexibility and/or envelope founded in robotic architecture alongside with the rigidity and efficiency provided by the machine tools kinematics. With the help of modern techniques that are used worldwide within different applications in various fields of the industries, and it is to be studied in the early study years of Bachelor of Science the Computer Aided Design (CAD) systems proved to be the correct solution for modeling-simulate-optimize before implement the final product. In this direction the work conducted through the research of this paper uses a modern, efficient, and adequate CAD system - SolidWorks. The current research focuses on using geometrical methods of identification and evaluation of the volume for the workspace of a 6 DOF PKM.

Keywords: Workspace analysis · PKM · CAD · Geometrical method · SolidWorks

1 Introduction

1.1 Importance of the Workspace of Parallel Kinematic Machines

Through the entire kinematics of these types of machines, PKM, have been proved to be a complex architecture, as it is to be found in the specific literature which underlines aspect that refers to the ability with high precision of changing the position and the orientation of these types of structures. In various domains such as microtechnology's or assembly electronic components and many others where are to be found robots based on a serial structure it has been proved that more suitable are the parallel kinematics machines due to the built-in repeatability, rigidity, and the accuracy of positioning [1–3]. Moreover, when it comes to reducing forces within the entire kinematic structure, PKM has proved to be efficient due to fixed frame fitting of the motors directly on the robot frame resulting a lower mass that needs to be moved [4, 5].

In terms of parallel robotic structures there are several research that have been developed to identify, measure, and analyze by any means using modern techniques the

B. Gapiński et al. (Eds.): MANUFACTURING 2022, LNME, pp. 50–63, 2022.
https://doi.org/10.1007/978-3-031-00805-4_5

workspace. It is to be considered that the workspace of a parallel robotic structure is a main characteristic, and it refers to it is performance [6, 7].

It is identified in the literature as an important aspect the aspect regarding the determination of this workspace and depending on the application for which the structure is intended to be implemented. With complex kinematic structures, PKMs are challenging structures in terms of workspace [8–13].

From the point of view of the applicability of these parallel kinematic structures, in the literature it is underlined the fact that the most used structure in this category, from a commercial point of view, having different dimensions, is the Gough-Stewart structure [14].

Also, it is underlining the aspect that refers to some limitation that PKM structure have, due to the workspace dimension. Workspace and its limitation are identified through its kinematic architecture which presents a fixed location of both active joints and passive joint within the base frame on which the arms are connected [1].

This paper presents workspace analysis and determination of a novel 6 DOF Parallel Kinematic Machine (PKM) using CAD in combination with evolutionary and numerical analysis methods.

Workspace determination is a critical issue as it measures the ability of the robot to work in a certain workspace due to the range of the movements of the robot arms. There are several ways of determining the robot workspace most of them are geometrical, numerical, and analytical [15–17]. Geometrical methods (CAD based methods) are preferred when cross sections of workspace are needed, numerical are preferred to show visibility and variation of several indices within the workspace. A combination of both numerical and geometrical method is preferred for workspace analysis and representation of limits of workspace [18, 19] (Fig. 1).

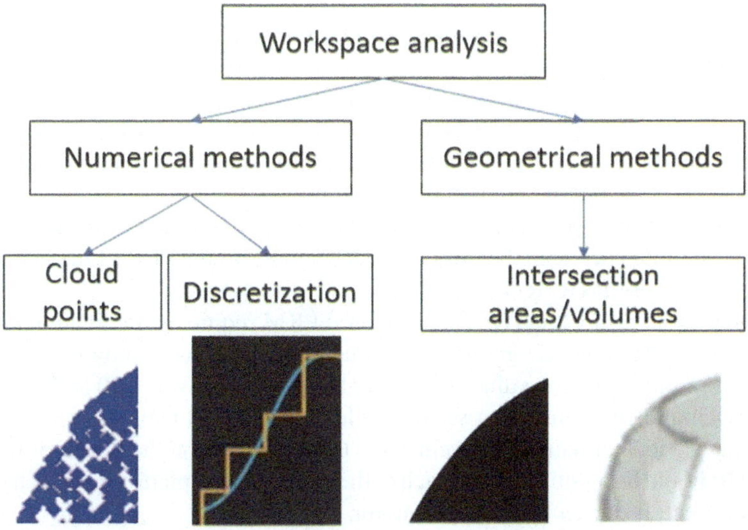

Fig. 1. Workspace analysis methods.

1.2 General Description of the PKM

The architecture of the novel concept of PKM type with 6DOF identified as 3-PR (RPR) RS has in its component two identical kinematic chains that are fixed on a mobile platform. Each kinematic chain is actuated through belt driven actuation fixed on two rails identified as the mobile base.

To properly control the structure, it is required to determine the number of degrees of freedom which will determine the minimal number of required actuators.

By using the formula for determining the degree of freedom of a spatial structure [20, 21] results:

$$M = 6n - 5C_5 - 4C_4 - 3C_3 - 2C_2 - C_1 \tag{1}$$

where:

- M - the degree of freedom of the studied structure
- n - the number of mobile elements
- C_n - the number of joints with (6-n) degree of freedom

In the PKM novel structure presented in this paper, the parameters described are:

- $n = 10$;
- $C_5 = 9$;
- $C_4 = 0$;
- $C_3 = 3$;
- $C_2 = 0$;
- $C_1 = 0$.

Through all mentioned above is can be underlining that the degree of freedom of the PKM architecture is 6. From the previous result it is understood the fact that the PKM structure requires at least 6 actuators to be completely controlled.

The joints that are going to be actuated will be the three prismatic joints of each kinematic chain by using a fixed belt drive with three stepper motors mounted on each prismatic joint and actuating the belt by using pinion.

The remaining three joints that need to be actuated are introduced as a new kinematic chain RPR, mounted between the links of the RRS joints of the PRRS kinematics chains, thus resulting the 3-PR (RPR) RS structure.

Kinematic chains no.1 and 3 of the proposed PKM require an inclination around the Z-axis to fully determine the geometrical structure. In the case study of the presented structure from this paper the values of the angles are respectively 120 and 60°.

The kinematic chain number two, the middle one, has an inclination of 60° around Y-axis, because its inclination determines the travel interval on the Z axis of the mobile platform. It should be mentioned that each of the three moving platforms, which represent a prismatic joint, is driven by a stepper motor.

All the prismatic joints that are constituent of the entire robotic structure PKM, are actuated to fully control the structure while the revolute and spherical joints are passive (Fig. 2).

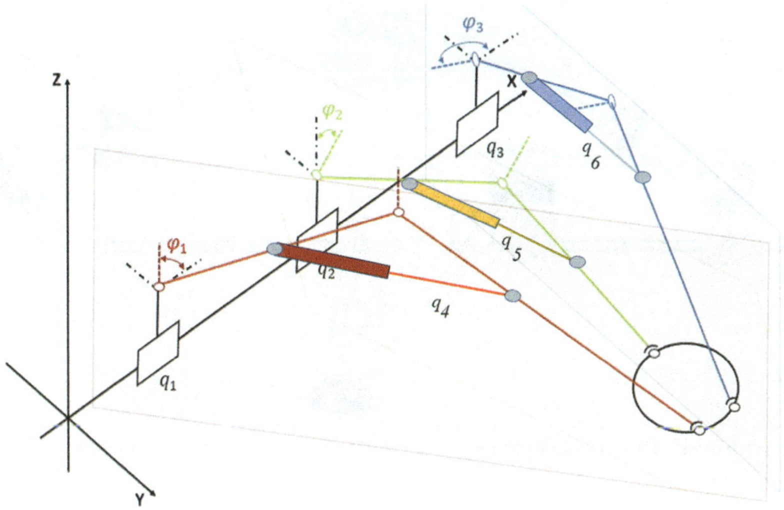

Fig. 2. Concept sketch of novel 6 DOF Parallel Kinematic Machine.

1.3 CAD Prototype of PKM

Following the general kinematic description of the entire architecture of the 6-DOF PKM, further it was used a CAD software application in order to define all the needed components which together mounted into an assembly design construction outlines the final form, see Fig. 3 and Fig. 4.

Within the design process it was taken into consideration even the next step which needs to be fulfilled, the manufacturing. In this direction in the 3D working environment all the parts were assigned with specific material, aluminum, and plastic. Authors have decided to combine two technologies of manufacturing, classic - lathe and milling together with non-conventional manufacturing technology (additive manufacturing) - 3D printing. It was established this to reduce the production cost regarding the parts manufacturing due to the fact of the high complexity.

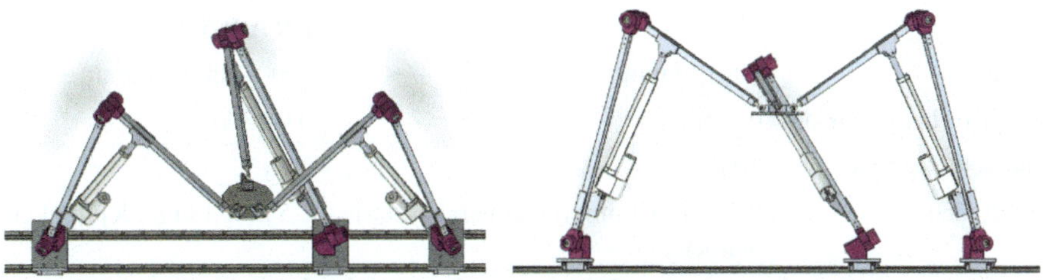

Fig. 3. Front and back 3D view of the novel 6 DOF Parallel Kinematic Machine.

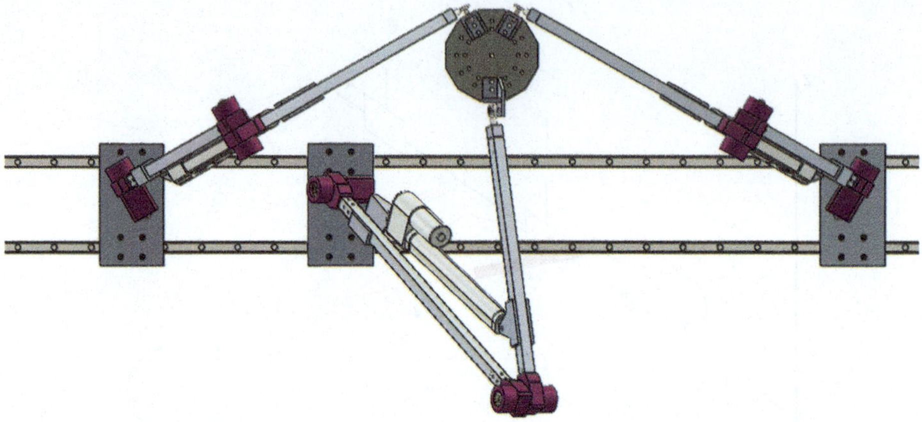

Fig. 4. Top 3D view of the novel 6 DOF Parallel Kinematic Machine

1.4 PKM Manufacturing and Assembly

The equipment's that were used to manufacture the entire parts of the structure assumed a 3D printer, Anet ET4 Pro, and a VMC, Microcut Challenger 2418, and the manufacturing parameters are presented in Table 1. Having the 3D models of the parts, the preliminary stage before the actual processing involved the use of the CAM (Computer Aided Manufacturing) environment to obtain the necessary information transmitted to the production equipment.

This stage was successfully completed with no errors or incompatibilities between the transfer of data from the CAD environment to CAM. The final parts obtained can be seen in the figures below Fig. 5, Fig. 6, and Fig. 7. In the pictures it can easily see the difference between the plastic parts, made on a 3D printer and those that have been processed by milling.

Table 1. Parameters used for manufacturing the parts

Parameter	Anet ET4 Pro	VMC Challenger 2418
Material used	PLA	EN AW-5083
Feed rate	20–100 mm/min	1000–3000 mm/min
Manufacturing precision	0.1 mm	0.01/300
Manufacturing time	80 h	20 h
Axial/Radial	0.2 mm/0.4 mm (layer height/nozzle diameter)	2.5 mm/3 mm (depth of cut)

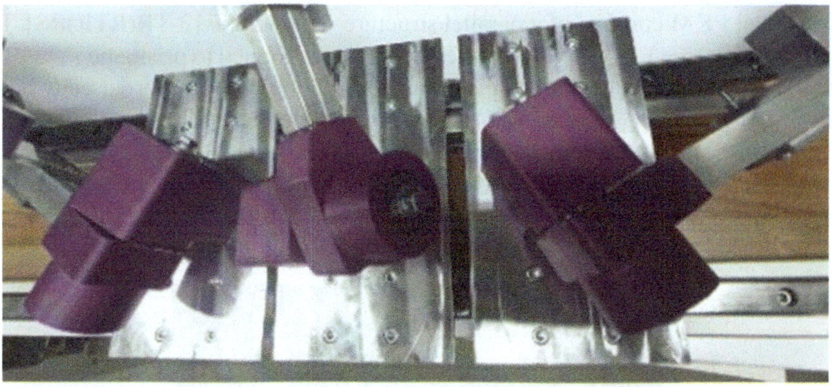

Fig. 5. Three prismatic joints and the first revolute joint of each kinematic chain mounted on the guiderails

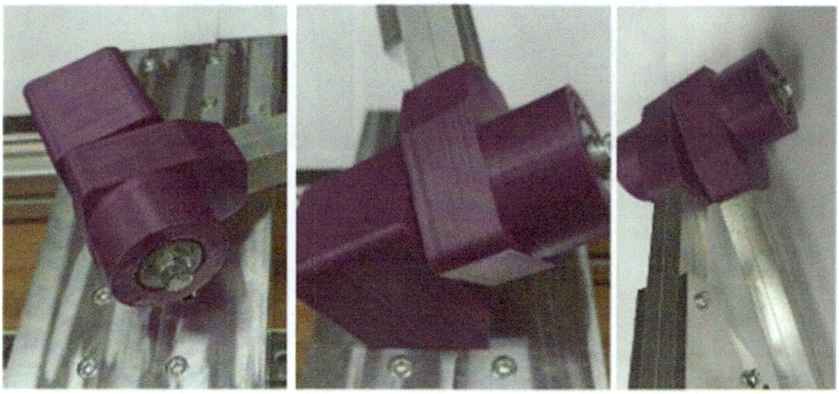

Fig. 6. Plastic elements mounted on the mobile platforms

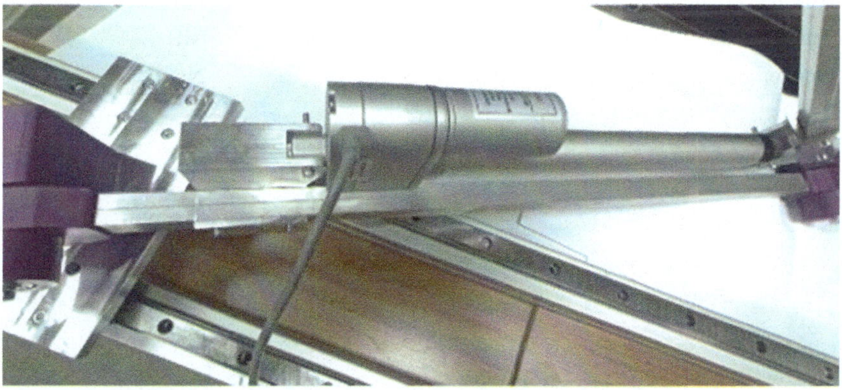

Fig. 7. Complete mounted arm having all the components assembled

The novel PKM consists of a parallel structure with 6 DOF [3-TR(RTR)RS]. It has 3 stepper motors, see Fig. 8 and parameters in Table 2 assigned to prismatic joints. From the point of view of actuation within the structure, stepper motors were used to achieve prismatic movements. A drive motor was used for each revolution torque.

A LACT10P linear actuator, see Fig. 8 and parameters in Table 3, with a stroke of 10 inches was used for each kinematic chain to perform the RRS type movements within the 3-TR (RTR) RS structure.

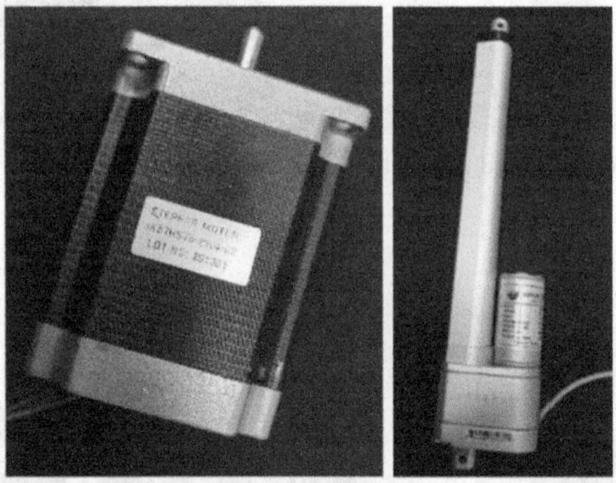

Fig. 8. Stepper motor and linear actuator used for actuating the structure

Table 2. Parameters of the stepper motors

Parameter	Value	Units
Model number	JK57HSB	
Current/Phase	1	A
Step Angle	1.8	°
Motor Height	86	mm
Inductance/Phase	0.9	MH
Holding Torque	0.39	Nm

Table 3. Parameters of the linear actuators

Parameter	Value	Units	Value [IS]	Units [IS]
Part number	LACT10P			
Stroke	10	Inch	254	mm
Input	12	V - DC	12	V - DC
Average speed	0.5	Inch/ s	12.7	mm/s
Dynamic load	110	lbs	49.9	Kg
Static load	500	lbs	226.8	Kg

2 Workspace Analysis

The determination of the workspace for any moving structure is an initial requirement for any further analysis involving the moving behavior of the robot with respect to a desired task. The determination of the workspace can be achieved by using geometrical, numerical, or analytical approaches and each method has its own advantages and disadvantages based on the desired workspace parameters that needs to be determined [22, 23].

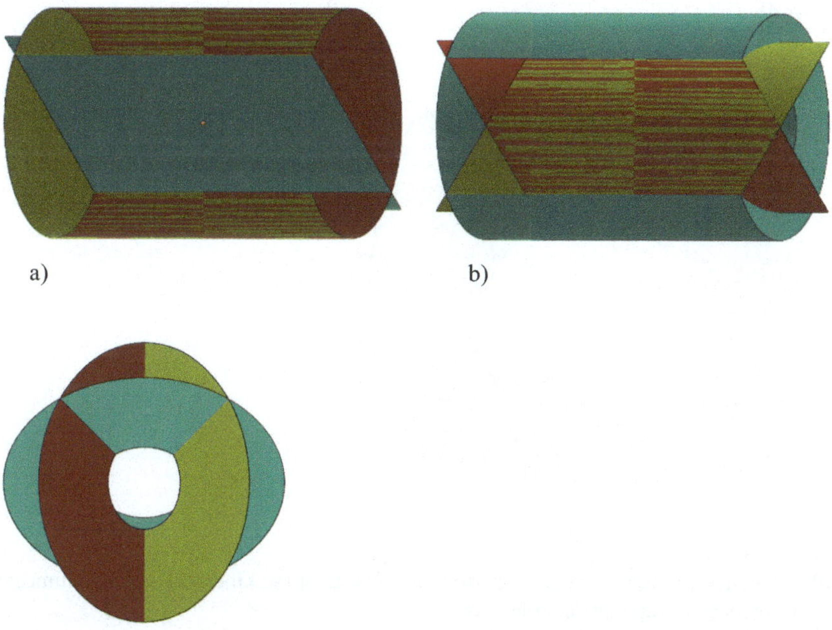

a) b)

c)

Fig. 9. Volumes generated by the three kinematic chains of the PKM a) front view, b) top view c) side view

Each approach used for the determination of the workspace can be implemented in multiple ways and based on implementation and precision factors some methods can be implemented with less difficulty than other. The geometrical method involves the procedures of generating the workspace volume by constructing the volume generated by each kinematic chain of the PKM and intersecting them, Fig. 9, thus resulting the workspace volume, Fig. 10, of the whole structure of the PKM. This method can be easily implemented by using CAD software and CAD tools and it will also result in the representation of the whole theoretical workspace. One downside of using this method in the case of PKM is the fact that certain construction parameters can be difficult to implement, like limit angles for the revolute joints, and like most PKM, the moving platform being the moving element that closes all the kinematic chains, is difficult to implement apart from the kinematic chains.

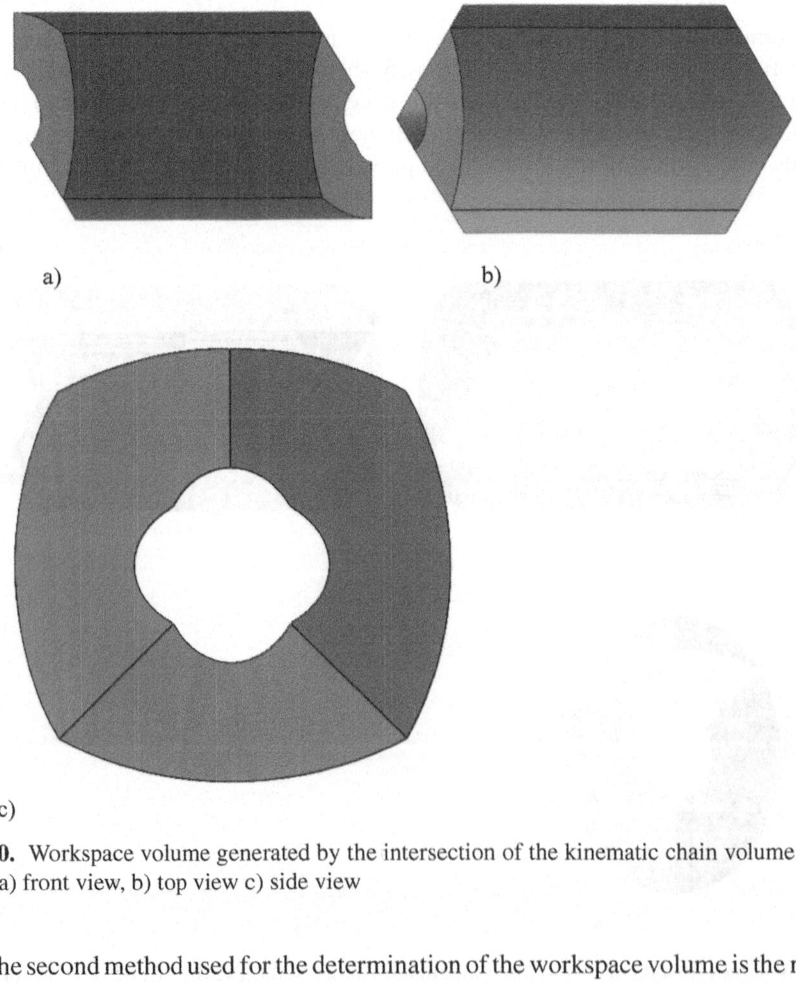

Fig. 10. Workspace volume generated by the intersection of the kinematic chain volumes of the PKM a) front view, b) top view c) side view

The second method used for the determination of the workspace volume is the numerical approach, which requires at least the equations that solve the forward kinematics problem or the inverse kinematics problem. By using the equations that solve the forward kinematics problem, a vector for each generalized coordinate is created and by iterating through each combination of generalized coordinates the result will be checked to conclude in a real position in space and if this is the case, that resulting point and orientation is stored and represented, thus resulting the reachable workspace [24].

By using the equations that solve the inverse kinematics problem, a grid of point is generated using fixed or variable space step and for each individual point and orientation the generalized coordinates of the PKM are computed and checked to result in real values of actuations. In this case, the initial point used for the determination is saved and represented, thus resulting the same reachable workspace.

The numerical method will only result in approximate representations of the workspace due to the implementation through iteration and the precision can be improved by decreasing the step used in the generation of initial generalized coordinates or positions and the construction parameters can be implemented with less difficulty (Fig. 11).

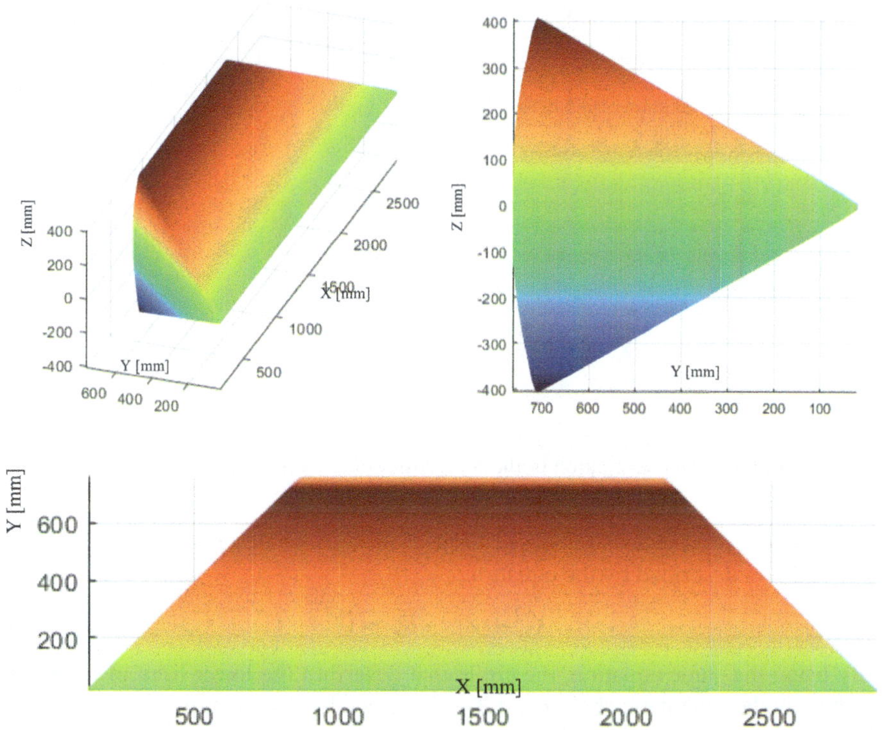

Fig. 11. Reachable workspace volume generated by numerical methods

In the case of the analytical determination of the volume, the approach is to determine the resulting surfaces from the interstation of the generated surfaces of each kinematic chain of the PKM and, through integration, to compute the resulting volume.

By using this method, the absolute value of the volume of the workspace can be computed and by further bounding the generated surfaces by using the limit angle for each joint, the absolute value of the reachable workspace can be computed [19, 25].

The numerical method will always result in the absolute values of the volumes for the workspace but in some cases, it is difficult to implement due to geometrical complexity.

For the implementation of the analytical approach, it is required to determine the functions that describe the projections of the shapes described by each kinematic chain on the YOZ plane, step followed by the determination of the shape on the intersection and finally the integration among X between the boundaries defined by the support planes of each kinematic chain (Fig. 12).

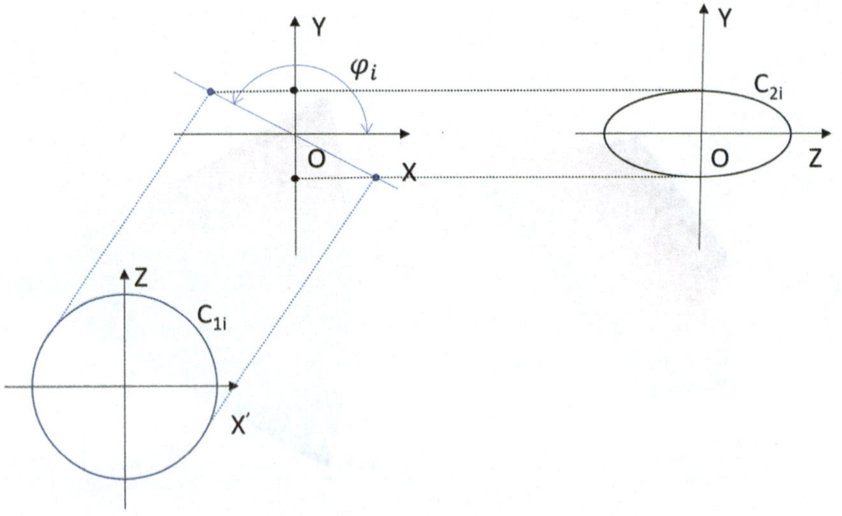

Fig. 12. Representation of the projections on the reference plane YOZ

The circle C_{1i} represents the upper limit of the surface described by the kinematic chain i and it is described by the equation:

$$X'^2 + Z^2 - R_i^2 = 0 \tag{2}$$

By using the same equation, the circle that describes the lower limit of the surface described by the kinematic chain i and it is described by the equation:

$$X'^2 + Z^2 - r_i^2 = 0 \tag{3}$$

Given the fact that the support plane for the circles is tilted around Z axis or Y axis by angle φ_i, the projection of the circles on the plane YOZ is:

$$\begin{cases} \frac{Y^2}{\sin(\varphi_1)} + Z^2 - R_1^2 = 0, \text{ outer circle} \\ \frac{Y^2}{\sin(\varphi_1)} + Z^2 - r_1^2 = 0, \text{ inner circle} \end{cases}$$

$$\begin{cases} Y^2 + \frac{Z^2}{\sin(\varphi_2)} - R_2^2 = 0, & outer\ circle \\ Y^2 + \frac{Z^2}{\sin(\varphi_2)} - r_2^2 = 0, & inner\ circle \end{cases}$$

$$\begin{cases} \frac{Y^2}{\sin(\varphi_3)} + Z^2 - R_3^2 = 0, & outer\ circle \\ \frac{Y^2}{\sin(\varphi_3)} + Z^2 - r_3^2 = 0, & inner\ circle \end{cases}$$

The boundaries of the workspace are the planes placed at the rear limits of the X axis oriented by the initial tilt angles and the whole volume is determined by:

$$V_{WS} = V_o - V_i, \tag{4}$$

where:

- V_{WS} - workspace volume,
- V_o - outer volume,
- V_i - inner volume.

The outer volume can be determined as:

$$V_o = \int S_0 dx \tag{5}$$

Where S_0 represents the function with respect to x of the section of the workspace.

The inner volume can be determined as:

$$V_i = \int S_i dx \tag{6}$$

Where S_i represents the function with respect to x of the section of the central workspace that needs to be removed.

Therefore, the workspace volume is determined by:

$$V_{WS} = \int S_0 dx - \int S_i dx = \int (S_0 - S_i) dx \tag{7}$$

3 Conclusions

Workspace determination of a parallel robot is a critical issue. Modern CAD software offer lately the possibility of representation complex shapes of intersected volumes and via constraints using evolutionary methods [27]. The article presented a concept of evolutionary design of the workspace volume of a novel 6 DOF PKM using SolidWorks. The proposed method consists in representation of the workspace as a volume in a CAD system (SolidWorks), based on CAD features like parametrization of it and we demonstrated the advanced performance of this proposed method to obtained volume of workspace of the PKM. The initial step relies on the parametric description of the workspace volume of the novel 6 DOF PKM and directly representation of it in the CAD system (SolidWorks). In the analysis process, design parameters were set, target value e.g., volume of workspace was obtained.

References

1. Palpacelli, M.C., Carbonari, L., Palmieri, G., D'Anca, F., Landini, E., Giorgi, G.: Functional Design of a 6-DOF Platform for Micro-Positioning. Robotics **9**(99) (2020)
2. Shimizu, Y., et al.: Design, and construction of the motion mechanism of an XY micro-stage for precision positioning. Sens. Actuators A Phys. **201**, 395–406 (2013)
3. Ahn, C., Seo, T., Kim, J., Kim, T.-W.: High-tilt parallel positioning mechanism devel-opment and cutter path simulation for laser micro-machining. Comput. Aided Des. **39**, 218–228 (2007)
4. Palmieri, G., Callegari, M., Carbonari, L., Palpacelli, M.C.: Mechanical design of a mini pointing device for a robotic assembly cell. Meccanica **50**(7), 1895–1908 (2015). https://doi.org/10.1007/s11012-015-0132-1
5. Righettini, P., Strada, R., Cortinovis, F.: Modal kinematic analysis of a parallel kinematic robot with low-stiffness transmissions. Robotics **10**, 132 (2021)
6. Arrouk, K.A., Bouzgarrou, B.C., Gogu, G.: CAD based techniques for workspace analysis and representation of the 3CRS parallel manipulator. In:19th International Workshop on Robotics in Alpe-Adria-Danube Region – RAAD 2010 Budapest, Hungary, 23–25 June 2010)
7. Masory, O., Wang, J.: Workspace evaluation of Stewart platforms. ASME Robot. Spatial Mech. Mech. Syst. **45**, 337–346 (1992)
8. Bonev, I.A., Ryu, J.: A geometrical method for computing the constant-orientation workspace of 6-PRRS parallel manipulators. J. Mech. Mach. Theory **36**, 1–13 (2001)
9. Pusey, J., Fattah, A., Agrawal, S., Messina, E.: Design and workspace analysis of a 6–6 cable-suspended parallel robot. J. Mech. Mach. Theory **39**, 761–778 (2004)
10. Gao, Z., Zhang, D.: Workspace representation and optimization of a novel parallel mechanism with three-degrees-of-freedom. Sustainability **3**(11), 2217–2228 (2011)
11. Guo, J., Wang, D., Fan, R., Chen, W., Zhao, G.: Kinematic calibration and error com-pensation of a hexaglide parallel manipulator. Proc. Inst. Mech. Eng. Part B J. Eng. Manuf. **233**, 215–225 (2019)
12. Nabavi, S.N., Shariatee, M., Enferadi, J., Akbarzadeh, A.: Parametric design and multi-objective optimization of a general 6-PUS parallel manipulator. Mech. Mach. Theory **152**, 103913 (2020)
13. Dong, W., Du, Z., Xiao, Y., Chen, X.: Development of a parallel kinematic motion simulator platform. Mechatronics **23**, 154–161 (2013)
14. Lin, L.-C., Tsay, M.-U.: Modeling and control of micropositioning systems using Stewart platforms. J. Robot. Syst. **17**, 17–52 (2000)
15. Stan, S.-D., Maties, V., Balan, R.: Workspace optimal design of a 2 DOF micro parallel robot using genetic algorithms and simulated annealing optimization methods. In: 2007 International Conference on Mechatronics and Automation, pp. 1108–1113 (2007). https://doi.org/10.1109/ICMA.2007.4303703
16. Merlet, J.P.: Determination of the orientation workspace of parallel manipulators. J. Intell. Robot. Syst. **13**, 143–160 (1995)
17. Agrawal, S.K.: Workspace boundaries of in-parallel manipulator systems. IEEE Trans. Robot. Autom. **7**(2), 94–99 (1991)
18. Arrouk, K.A., Bouzgarrou, B.C., Stan, S.D., Gogu, G.: CAD based design optimization of planar parallel manipulators. In: Solid State Phenomena, vol. 166–167, pp. 33–38. Trans Tech Publications, Ltd. (2010). https://doi.org/10.4028/www.scientific.net/ssp.166-167.33
19. Arrouk, K.A., Bouzgarrou, B.C., Gogu, G.: Workspace characterization and kinematic analysis of general spherical parallel manipulators revisited via graphical based approaches, Mech. Mach. Theory **122**, 404–431 (2018). ISSN 0094-114X, https://doi.org/10.1016/j.mechmachtheory.2017.11.019

20. Han, H., Zhang, Y., Zhang, H., Han, C., Li, A., Xu, Z.: Kinematic analysis and performance test of a 6-DOF parallel platform with dense ball shafting as a revolute joint. Appl. Sci. **11**, 6268 (2021). https://doi.org/10.3390/app11146268
21. Sreenivasan, S.V., Waldron, K.J., Nanua, P.: Closed-form direct displacement analysis of a 6–6 Stewart platform. Mech. Mach. Theory **29**, 855–864 (1994)
22. Chaudhury, A.N., Ghosal, A.: Determination of workspace volume of parallel manipulators using Monte Carlo method. In: Zeghloul, S., Romdhane, L., Laribi, M.A. (eds.) Computational Kinematics. MMS, vol. 50, pp. 323–330. Springer, Cham (2018). https://doi.org/10.1007/978-3-319-60867-9_37
23. Gokul Narasimhan, S., Shrivatsan, R., Venkatasubramanian, K., Dash, A.K.: Determination of constant orientation workspace of a Stewart platform by geometrical method. Appl. Mech. Mater. **813–814**, 997–1001 (2015). https://doi.org/10.4028/www.scientific.net/AMM.813-814.997
24. Abdel-Malek, K., Yeh, H.-J., Inc, M., Khairallah, N.: Workspace, void, and volume determination of the general 5DOF manipulator. Mech. Struct. Mach. **27** (2000). https://doi.org/10.1080/08905459908915690
25. Bonev, I., Gosselin, C.: Analytical determination of the workspace of symmetrical spherical parallel mechanisms. Robot. IEEE Trans. **22**, 1011–1017 (2006). https://doi.org/10.1109/TRO.2006.878983
26. Konig, O., Wintermantel, M.: CAD-based Evolutionary Design Optimization with CATIA V5. Weimarer Optimierungs- und Stochastiktage 1.0, 2/3. December 2004
27. Oarcea, A., Popister, F., Stan, S.D., Cobilean, V.: Comparative study of CAD optimization features for the workspace of 3DOF Parallel Robot. In: 2021 9th International Conference on Modern Power Systems (MPS), pp. 1–6 (2021). https://doi.org/10.1109/MPS52805.2021.9492529

Development of the Globelike Mobile Robot with the Manipulator Set

Jędrzej Urbański(✉), Mateusz Stachowiak, and Dominik Rybarczyk

Poznan University of Technology, Plac Marii Skłodowskiej-Curie 5, 60-965 Poznań, Poland
jedrzej.urbanski@student.put.poznan.pl,
dominik.rybarczyk@put.poznan.pl

Abstract. The article presents the concept of an innovative mobile robot with a set of manipulation arms. The peculiarity of the device is the way of moving in every direction without having to change the orientation of the structure. This effect was achieved thanks to the spherical shape of the body. A set of arms cooperating with it are used to manipulate objects within the operating range of the device. The control system uses microcontrollers, orientation and position sensors. The robot is controlled remotely by the user using a smartphone. Based on the literature review, no similar solution used in the industry was found.

Keywords: Mobile robots · Ball robot · Robot manipulator · Control system

1 Introduction

1.1 Review of the Literature

In large-scale production plants whose factories occupy huge spaces in production halls, inter-station transport is very important. It is more and more common for enterprises to invest in mobile robots, the task of which is to transfer elements from one station to another. They usually look like a wheeled platform that follows a certain, predefined trajectory. Thanks to this, employees can focus only on their work, not paying attention to the need for the necessary components. This solution also has the advantage that the transport robot is reliable. There are accidents where a forklift operator, through excessive fatigue or distraction, causes large material losses in the company or, worse, injures other people. The use of robots avoids this type of accident [1].

Due to the way robots move in space, they can be divided into holonomic and nonholonomic. Each mobile robot has generalized coordinates q and generalized speeds. The relations between the robotic system and its environment are described by the configuration constraints $f(q) = 0$ and the phase constraints $A(q) = 0$. Phase constraints may be possible to integrate - then they are holonomic phase constraints or impossible to integrate - nonholonomic phase constraints. Therefore, holonomic robotic systems are those whose constraint equations can be integrated, while if the equations are incomplete, we are talking about nonholonomic systems.

Another, more intuitive way to distinguish between robots as holonomic and non-holonomic involves the number of degrees of freedom of the robot and the number of

B. Gapiński et al. (Eds.): MANUFACTURING 2022, LNME, pp. 64–77, 2022.
https://doi.org/10.1007/978-3-031-00805-4_6

degrees of freedom that the robot is able to control. If these numbers are equal, then we are talking about a holonomic work, if not - a nonholonomic one.

Among mobile, wheeled robots, a significant part is occupied by non-holonomic robots. The main source of nonholonomic limitations is the lack of wheel slip - the wheel speeds perpendicular to the wheel cancels out. The solution to this problem may be the use of omnidirectional wheels, also known as Swedish wheels. These wheels, on their circumference, have rollers enabling such a wheel to be moved in a direction perpendicular to the normal direction of wheel movement [2].

In certain cases, it is required to use a manipulator, which will additionally have the functionality of a mobile robot. Such a need occurs, for example, in the military industry, namely in the performance of activities that are dangerous to humans, such as defusing bombs or reconnaissance. Another application in which it is necessary to be able to move as well as perform manipulator operations is the space industry. In this case, an additional important element is the control, which must be reliable.

The paper proposes the construction of a sphere-shaped robot [3]. Ultimately, this robot is to work with a set of arms mounted magnetically (with magnets) on the main body. As far as it is known such or similar combination has not been introduced yet. The goal of this research is to check whether such solution makes sense. Most mobile robots with manipulators e.g., Kuka KMR Quantec [4] or Omron Moma [5], consist of a 4-wheel platform equipped with one manipulator. The proposed, spherical robot kinematics is characterized by one point of contact with the ground, which ensures greater movement capabilities of the robot in unpredicted or demanding terrain. Two manipulation arms instead of one provide easier balancing of the center of gravity of the robot holding a carried element. The main purpose of the described contraption is the safe transport of goods in the warehouses of different kind. The selected kinematics does not have high positioning accuracy, while the spherical shape of the carrier part ensures that the robot cannot be overturned and the transported goods stay intact. Manipulators facilitate the maintenance of balance and allow for arranging the transferred elements.

The project covers the mechanical structure, power supply system, drive system, control system and the method of communication between a human and a robot. The robot has the shape of a sphere with a diameter of approx. 30 cm, is controlled wirelessly using an Android application, can move in any direction on the plane and has an "orientation" function that allows the user to easily determine the direction of movement.

1.2 Theory of Operation

Although the robot has the shape of a sphere, its motion can be considered as a disk [6], as shown in Fig. 3. The center of mass of IDU (Internal Driving Unit) M, shifted in relation to the geometric center, is affected by the force of gravity m. This force on the arm l in relation to the instantaneous center of rotation S causes a driving torque τ.

$$\tau = l \times m = rmsin\theta \quad (1)$$

The torque can also be expressed as the product of the moment of inertia I and the angular acceleration α.

$$\tau = I\alpha \quad (2)$$

As we know, giving the body an angular acceleration causes it to rotate, making the robot roll. The influence of the value of the angle θ on the value of the torque τ is also important. The torque has the greatest value for $\theta = 90°$, i.e., in the position where the center of gravity of the IDU is at the greatest distance in the plane perpendicular to the direction of the force of gravity from the point of contact S. The mass of the IDU also directly influences the torque value. It follows that in order to obtain the greatest possible angular acceleration, the mass and the angle of deviation of the IDU should be as high as possible, and the moment of inertia of the robot - as small as possible (Fig. 1).

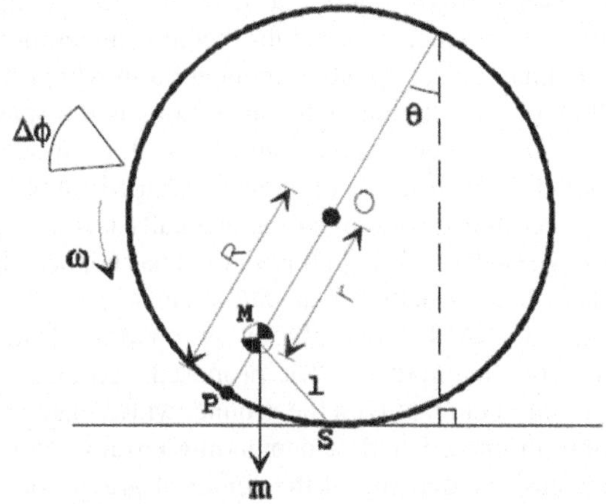

Fig. 1. Model of a robot as a disk. O - geometric center, S - point of contact with the ground, M - IDU center of mass, r - distance between the center of the ball and IDU, l - distance between the IDU and the contact point, θ- IDU deflection angle, ω - angular velocity of the ball, m - IDU gravity force [6]

The main robot movement is coming from the rotational speed of the drive wheel ωD. The purpose of controlling the speed ωD is to cause the angular acceleration α. The acceleration α can be given by changing the value of the angle θ, which can be changed by the motion of the IDU, i.e., making the wheel move at the speed ωD.

From the equality of linear velocities at the point of contact of the wheel with the ball:

$$\omega R_1 = \omega_D r_D \tag{3}$$

where:

R_1 - radius of a disk,

r_D - radius of a drive wheel.

which allows to derive a simple formula on the rotational speed of the ball:

$$\omega = \frac{r_D}{R_1}\omega_D \tag{4}$$

making it easy to control the speed of the ball with the speed of the drive wheel.

The problem in controlling the robot from the user's point of view is determining the direction in which the robot is facing, so it is not known which direction the user is pointing on the application screen. The solution to this problem uses data about the angle between the north direction and the direction indicated by the robot's characteristic point or the telephone, i.e., azimuth. The concept of this solution is presented in Fig. 4. In the control without "orientation", the user specifies the angle ρ in relation to the robot's characteristic point (front) using the indicating element (3 - analog knob), which determines the direction of the desired movement. When the user knows where the robot's front is located, he can intuitively adjust the angle of the analog stick to move the robot in the selected direction.

In the case of a spherical robot, it is impossible to define the front as the platform is inside a closed sphere. So, the problem is the difference in the directions indicated by the robot and the smartphone. Knowing these directions, it is possible to correct the angle of the knob set by the operator in such a way that the movement of the ball actually takes place in the direction specified in the application. This dependence was expressed by the formula:

$$\rho_{steering} = \rho + (\varphi - \gamma) \tag{5}$$

where:

$\rho_{sterring}$ – angle defining direction in which the robot should move in relation to its front,
ρ – angle between direction chose by user and direction pointed by smartphone,
φ – smartphone azimuth,
γ – robot azimuth (Fig. 2).

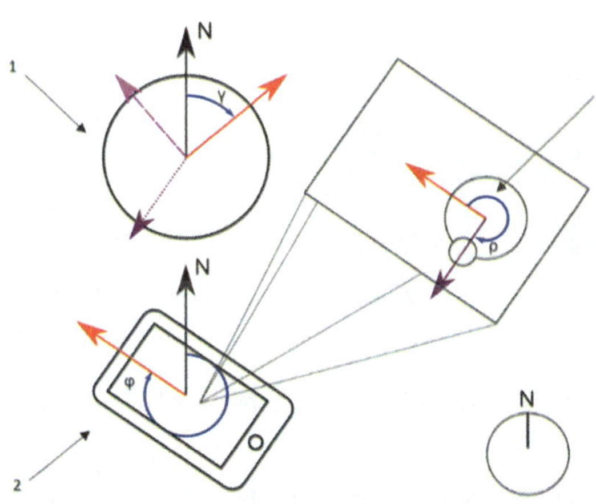

Fig. 2. Steering the direction of movement with "orientation" feature; 1 - robot; 2 - smartphone with running application; 3 - steering stick; purple, solid line arrow - direction pointed by user; pink, dashed line arrow - the direction in which robot will move with "orientation" off; purple, dotted line arrow - the direction in which robot will move with "orientation" on

2 Mechanical Construction

2.1 Driving Platform

The idea of the robot - the sphere determines its external shape. The robot consists of two parts - the outer shell and the platform inside. The ball must be rigid enough not to deform under the weight of the platform which could prevent effective movement. At the same time, it must not be brittle, it must have sufficient impact strength to withstand collisions resulting from the ball running into the surroundings. The ball may not be made of a material that blocks radio communication, e.g., metal, plastic is preferred. Due to the possibility of cooperation between the robot and the unit with the arms, the thickness of the shell cannot be too great for the magnetic force to be sufficient to hold the cap. The surface roughness is also an important aspect. Both the outer and inner surfaces of the shell must not be too smooth, which could interfere with movement.

The main and basic task of the internal platform is to provide the connection and arrangement of other elements such as motors, battery, sensors, control electronics, switches. The platform should be seated as low as possible in relation to the geometric center of the ball, thanks to which the amplitude of the ball deflection caused by external excitation will be reduced to a minimum. The bending strength of the wheel spacing members must be sufficient, i.e., withstanding the weight of the platform, as they support the entire structure. Due to the location inside the magnetometer sphere, there must not be any ferromagnetic or other material that could interfere with the operation of the sensor in its vicinity.

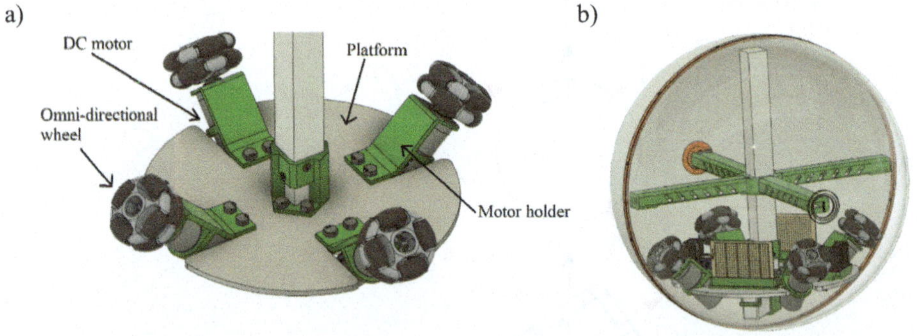

Fig. 3. a) view of the inner platform of the robot, b) view of the whole robot ball.

The challenge was to construct a separable shell. This problem was solved by using two hoops. They were made by the method of additive manufacturing in layers - FFF. The purpose of the rim is to connect two hemispheres. The hoops on the two hemispheres differ from each other and their shapes are designed in such a way that they fit into each other. In addition, the rims are screwed together with a bolt and a nut. In one of the variants of the rim, a nut is mounted. Then a second hoop is put on and both elements are screwed together with a screw. The screw used is a M1.6 x 2.5. clamping screw with an allen slot.

2.2 Grippers

Achieving the assumed object manipulation goal requires a mechanical interface that will allow the manipulator to cooperate with the mobile part of the robot, which is a rolling sphere. It should be emphasized that in this case it is impossible to attach the platform on which the arms will be mounted directly to the sphere, due to the fact that the platform with the set of arms must be able to slide freely on its surface. This is possible thanks to the mechanical connection in the form of neodymium magnets. The permanent magnetic field will not affect the operation of the robot, and will allow it to work with the ball. Tower Pro MG-996R modeling servos were used to implement the movement of the manipulation system arms. The problem of gripping objects was solved by using the set of Pololu - Micro Gripper Kit. Thanks to this gripper, it is possible to manipulate objects up to 32 mm in size, since the jaws of the component open to this width [7]. The gripper mechanism is based on the use of a toothed wheel and a toothed bar [8]. To avoid unnecessary friction of magnets or structural elements against the surface of the ball, 4 Pololu metal balls with a diameter of 3/8 inch, embedded in a dedicated holder, were used. Thanks to them, the structure is not supported on magnets, and this function is performed by balls, the so-called Ball Caster. This solution allows to obtain a lower value of kinetic friction due to the fact that the rolling friction coefficient is lower than that of the sliding friction. All these components are mounted on an additive manufacturing structure. For this purpose, PLA (polylactide) material was selected [9], which belongs to the group of thermoplastics. In the era of high generation of plastic waste, polylactide, compared to other materials of this type, is distinguished by biodegradability. Of course, this does not mean that an element made of PLA that will be thrown in a landfill will decompose on its own [8].

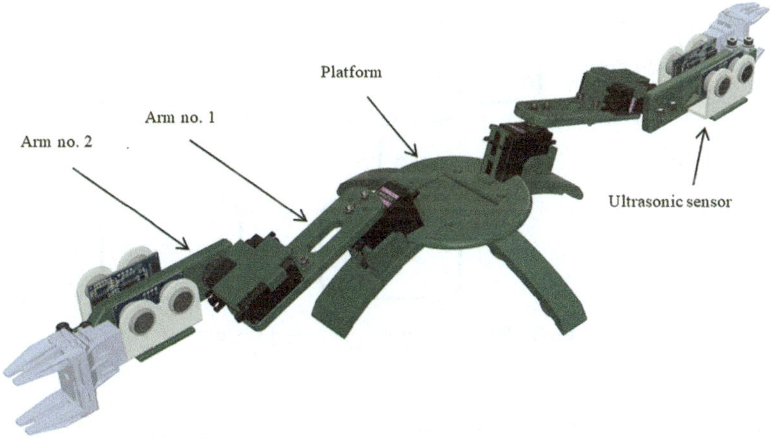

Fig. 4. Complete manipulation system

3 Control System

3.1 Driving Platform Control Hardware

The role of the electronic system in the robot is to collect data from sensors, process them according to a programmed algorithm and appropriate control of actuators. The program execution takes place in a microprocessor unit, most often a microcontroller. As in any system, power source is also needed, in the case of a mobile robot - a battery. Wireless communication must also be ensured in the designed device.

Expectations for motor controllers include the ability to change the direction of the motor rotation, the option of smooth speed control (for example by means of PWM [11]) and the ability to ensure sufficient current consumed by the motor. As a system of sensors, the IMU is preferred as an accelerometer, gyroscope and compass [12]. Wireless communication can be provided by means of systems working in Bluetooth or Wi-Fi standards, or by using an RF transmitter and receiver [13]. The first solution, however, will enable cooperation with other devices that support these standards, such as smartphones or PCs, thanks to which it will be possible to control the robot using a mobile application. For the operation of sensors and controllers and for the execution of the control algorithm, a microcontroller is preferred, much smaller in size and more energy-efficient than a mini-computer or a single-board computer (Fig. 5).

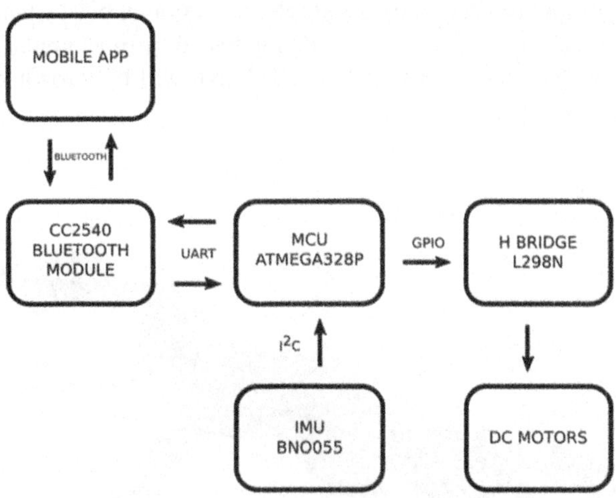

Fig. 5. Block diagram of control system.

3.2 Driving Platform Control Software

The task of the control application for mobile is primarily to retrieve input data from the user, retrieve information from the robot, process it and send control data back. The application performs the correction of the traffic direction presented in Sect. 1.2. In addition, it is also possible to control parameters such as switching on/off the motion, switching on/off the regulator, rotation around its own axis, etc.

The first function of the microcontroller software is collecting information from the orientation sensor via the I2C interface. Another one is carrying out the entire process of PID control with a constant frequency (calculation of errors, output values, etc. [14]), and then appropriate driving of the microcontroller outputs in order to control the motors. The microcontroller must also receive data sent from the phone, as well as send information to it, including readings from the inertial sensor. The information sent from the robot is primarily the orientation angles and, additionally, auxiliary messages informing about the robot's state. Data should be sent with a constant frequency. A useful function would also be to monitor whether the user is connected via the phone application to avoid unwanted behavior of the robot.

The microcontroller software is written in C/C++. The program runs continuously - it executes an infinite loop. With each run of the loop, a sequence is performed. It consists of the following tasks: receiving data sent from the application, processing the received data, sending information about the robot's status to the phone, making adjustments or direct actuation of the motors. These tasks are performed with different frequency, which means that not all of these functions will be performed in every cycle. Data acquisition and transmission is carried out by data frames that have start and stop characters at the beginning and end. The receiving function writes into the array only the data that is between these characters. Upon encountering a stop character, writing to the array is complete and the newData flag is set to true. The end time of data reception is also stored, which is used to detect connection interruption. If the time from the last data reading is too long than the fixed one, the table storing the received data is reset and the new data flag is set to true.

The robot control system can control the movement of the ball in two independent, perpendicular axes x and y. In order for the robot to move, two values X and Y must be sent from the mobile application, specifying the amount of movement in a given axis. Getting the robot to move in any direction is possible by adding the moves in both directions. For this purpose, it is necessary to convert the speed of any direction and value into the sum of two speeds in the x and y directions. The mobile application is responsible for this, thanks to which specific values for controlling the robot's axes of movement are transferred to the microcontroller program. Depending on the user's desire, the robot can be controlled with or without regulators.

To set the robot in motion, move the platform out of its equilibrium position - tilt it by a certain angle. For this purpose, the microcontroller program implements a regulator that's role is to maintain the deviation angle at the indicated level. There is a separate controller for each axis. The setpoints for the controllers are rescaled X and Y values received from the application. The deviation angle around the x axis is marked with the letter A, and around the y axis with the letter B. It means that changes in angle A cause movement in the y axis, and changes in angle B - movement in x axis. Applied controllers are PID controllers. In the actual implementation, the operation of the regulator is discrete and takes place from time to time, in this case at a frequency of 100 Hz (Fig. 6).

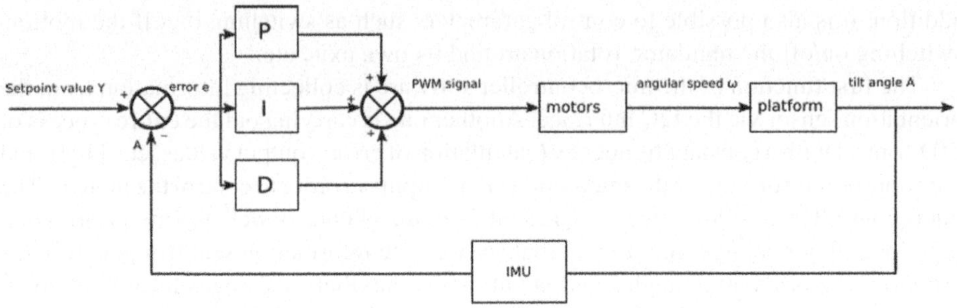

Fig. 6. Regulation system of tilt angle "A" using PID controller.

3.3 Grippers Control Software

The next positions of the servos are determined by the application for smartphones with Android software, written for the needs of the project. The data is sent via the Bluetooth module and received by the microcontroller. The data registered by the controller are appropriately converted, and then the desired outputs are activated. The process of moving the arms is supervised by ultrasonic sensors whose task is to prevent possible collisions of the manipulator arms with obstacles. The sensors influence the movement performed by the first modeling servos from the base.

If the control concerns second arm or gripper drives, the movement is performed immediately, while in the case of first arm servos, the signal from the ultrasonic sensors is checked before starting the movement. When an obstacle is detected closer than the set limit distance, the movement will be stopped, and if it was just about to start, the arm will not move at all. When the obstacle is on one side of the arm, there is no contraindication to move towards the side where the collision will not occur. This means that movement is blocked only in the direction of the obstacle and does not affect movement in other directions (Fig. 7).

The movement of individual arms is controlled by sliders. After changing their position, a sequence of characters is sent, with the appropriate prefix, so that there is no doubt which modeling servo should work at the moment. If the connection is not established, the items are not sent to avoid possible errors in the application. The program can be used in both the horizontal and vertical position of the screen. When the phone is flipped over, the application is scaled in width, however, it is necessary to scroll to access all functionalities. Due to the large number of elements arranged one above the other, it is recommended to use the application in a vertical orientation, so that all sliders and buttons are visible at all times, which significantly improves the comfort of use.

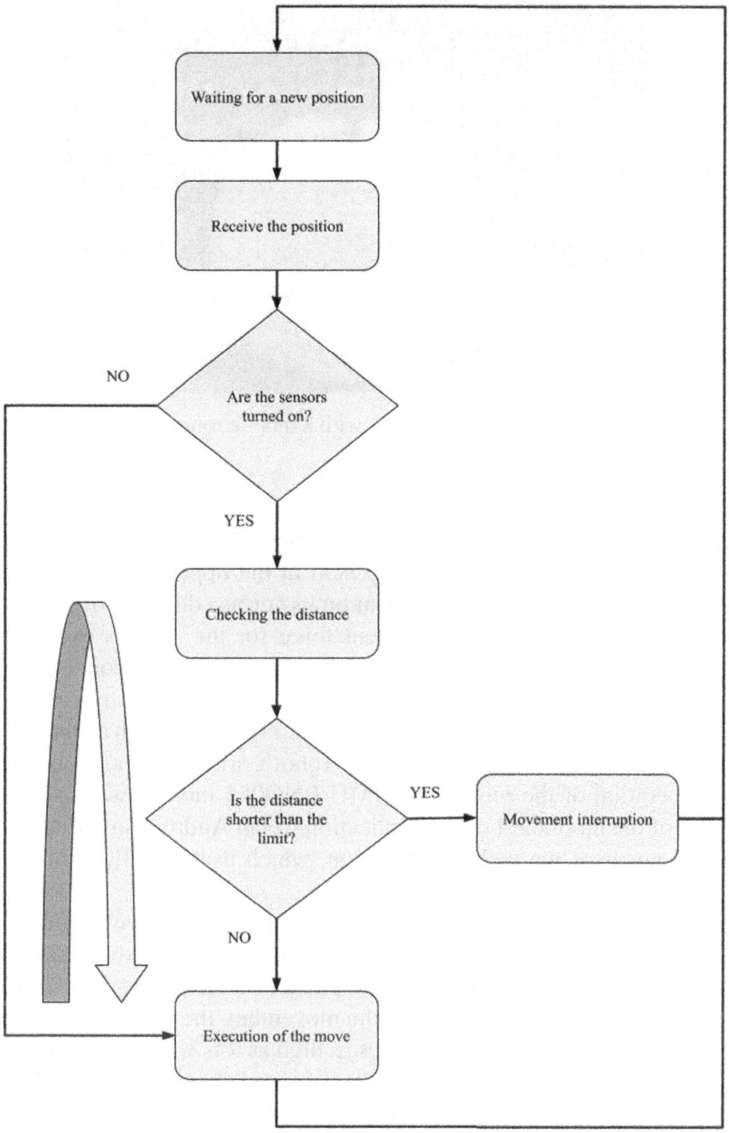

Fig. 7. Block diagram of the control program.

4 Experimental Tests and Discussion

The manipulator was made and tested, which confirmed that all assumptions were met. During the tests, there were no problems suggesting the need to modify the structure or the control system. Selected components, both mechanical and electronic, have proven successful in use, so you can confirm their correct selection (Fig. 8).

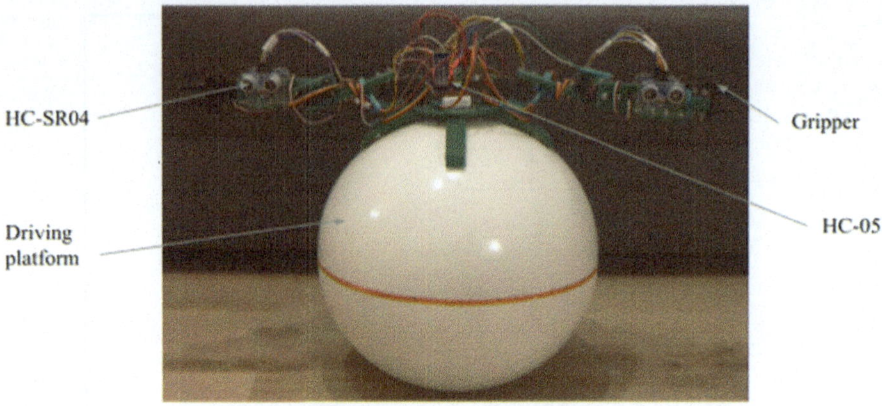

Fig. 8. Manipulator with a mobile robot.

The functionality of the manipulator was also checked in direct cooperation with the sphere-shaped robot for which it was designed. The used neodymium magnets ensured that the manipulation assembly was firmly held in the upper part of the ball, and the Ball Caster balls allowed for easy movement on its surface during movement. This type of detachable connection provides sufficient force for the devices to cooperate, but, if necessary, it does not cause difficulties in separating the components. The mobile robot is equipped with 4 DC motors that allow the assembly to move in any direction without the need to rotate. Its power source was a Li-Pol battery with an output voltage of 11.1 V. In order to obtain information about the robot's orientation, as well as to properly regulate the operation of the motors, the IMU BNO055 module was used. To control the movement of the mobile robot, an application for an Android smartphone was used, written for the needs of the technical solution, which uses the Bluetooth interface to communicate with the controller.

From the comparison of the graphs (Fig. 9, 10), it is easy to see that the regulator is working. Its action significantly reduces the time of the ball to stabilize after being out of equilibrium from about 40 s to 10. The amplitude of deflections is also significantly reduced. Already at the very beginning of the movement, the oscillations of the ball are smaller and are not even able to reach values as high as it is when the regulator is turned off. It is worth noting that the ball during the process behaves like a pendulum, which oscillations are damped. The controller helps to reduce these vibrations to a minimum in the shortest possible time.

In order to verify the correct operation of the device in terms of preventing the arm from hitting an obstacle, measurements were taken. For each of the four ultrasonic sensors, the value of the distance from the nearest obstacle was recorded for each position of the shaft Θ during the movement. The blue line on the graph is the distance from the obstacle, the horizontal red line is the value at which the movement should stop, and the green diamond shows the position at which the obstacle stopped the movement. The situation in which the blue line would be under the red line is unacceptable and would indicate a malfunction of the system (Fig. 11).

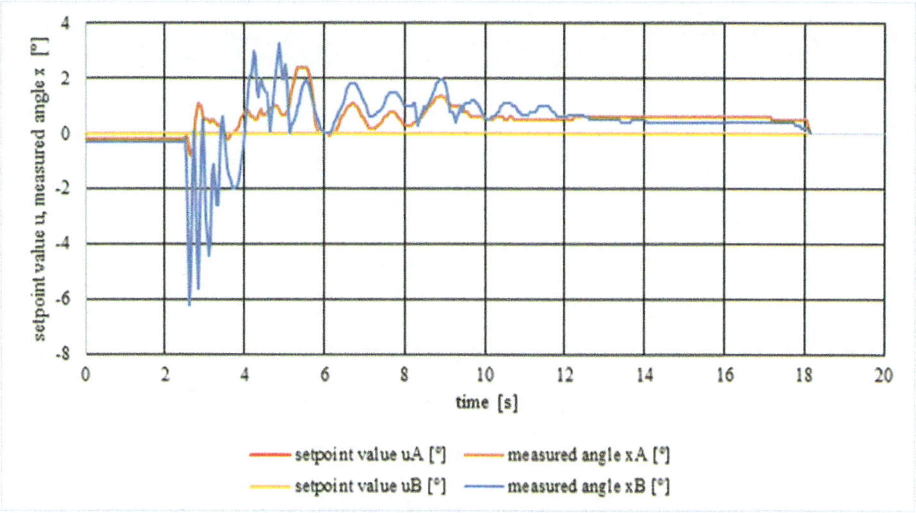

Fig. 9. Graph of the platform's inclination angle during disruption with the regulator turned on.

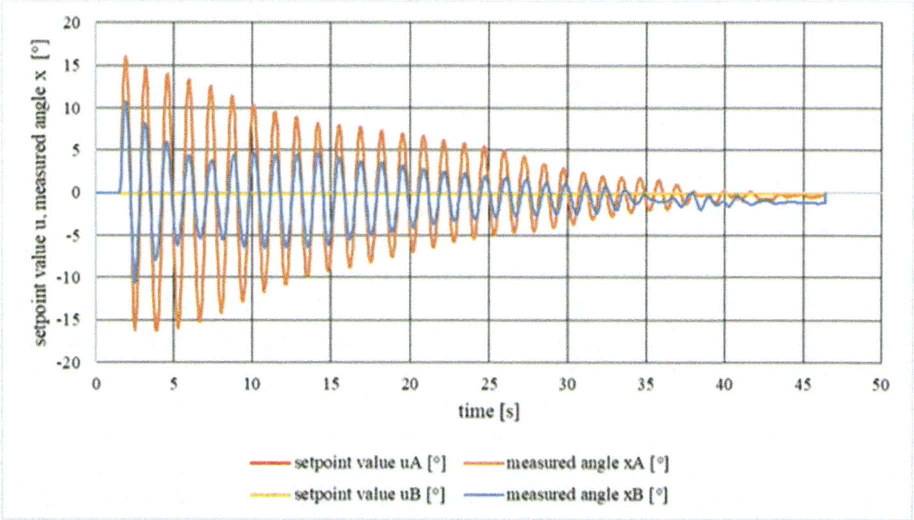

Fig. 10. Graph of the platform's inclination angle during disruption with the regulator turned off.

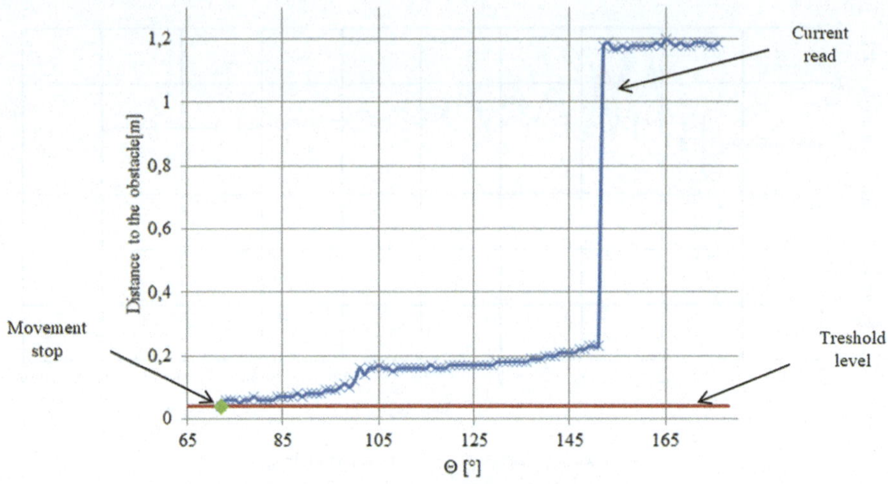

Fig. 11. An exemplary graph of the distance from the obstacle for the left arm

5 Conclusions

The conducted research proved that it is possible to build a robot with the presented kinematics. The big advantage of the constructed device is the robot's spontaneous return to the point of equilibrium. Thanks to this, the role of the control system is only to support the phenomenon that occurs naturally. The obtained results are satisfactory, considering the fact that low-specialized components were used. The limitations that have been encountered are, first of all, friction and slippage between individual components, play in mechanical transmissions, or poor quality of components. The direction of development of further research is the integration of the driving and handling parts in order to provide additional balance by means of a pair of arms.

The decision to prepare a smartphone application for a specific solution provided full freedom in controlling and arranging elements on the screen. It is true that ready-made applications from the "Play Store" could be used for control, but to some extent they could limit the functionality of the manipulation system, which would certainly reduce the transparency of the interface and the comfort of use. In addition, when developing an application with a specific application in mind, in the event of problems or possible software updates, there is no problem with making any modifications. The undeniable advantage of such a solution is also the simple way of implementing the application using a programming environment that can be used for free. Even people who have not had any contact with the implementation of projects regarding applications for mobile devices will certainly be able to make a basic application. Of course, basic programming knowledge is required for this, such as knowledge of conditional instructions. All the above-mentioned advantages mean that when implementing smartphone control into a project, it is worth considering preparing an application program tailored to the needs of the solution.

Acknowledgment. This research was funded by grant number 0614/SBAD/1547.

References

1. Rubio, F., Valero, F., Llopis-Albert, C.: A review of mobile robots: concepts, methods, theoretical framework, and applications. Int. J. Adv. Robot. Syst. **16**(2), 1729881419839596 (2019). https://doi.org/10.1177/1729881419839596
2. Practical Applications for Mobile Robots based on Mecanum Wheels–A Systematic Survey.pdf, pp. 21–29. https://www.incdmtm.ro/editura/documente/pag.%2021-29.%20Practical%20Applications%20for%20Mobile%20Robots%20based%20on%20Mecanum%20Wheels%20-%20A%20Systematic%20Survey.pdf. Accessed 24 Oct 2021
3. Kaznov, V., Bruhn, F., Samuelsson, P., Stenmark, L.: Ball robot. US8099189B2, 17 January 2012. https://patents.google.com/patent/US8099189B2/en. Accessed 25 Oct 2021
4. KMR QUANTEC. https://www.kuka.com/en-us/products/mobility/mobile-robot-systems/kmr-quantec. Accessed 14 Jan 2022
5. Omron Mobile Manipulator Solution. https://industrial.omron.eu/en/solutions/product-solutions/omron-mobile-manipulator-solution. Accessed 14 Jan 2022
6. Halme, A., Schonberg, T., Wang, Y.: Motion control of a spherical mobile robot. In: Proceedings of the 4th IEEE International Workshop Advances Motion Control–AMC 96–MIE (1996). https://doi.org/10.1109/AMC.1996.509415
7. Zaidi, S., Maselli, M., Laschi, C., Cianchetti, M.: Actuation technologies for soft robot grippers and manipulators: a review. Curr. Robot. Rep. **2**(3), 355–369 (2021). https://doi.org/10.1007/s43154-021-00054-5
8. Tai, K., El-Sayed, A.R., Shahriari, M., Biglarbegian, M., Mahmud, S.: State of the art robotic grippers and applicationsx. Robotics **5**(2) (2016). Art. no. 2, https://doi.org/10.3390/robotics5020011
9. https://pdfs.semanticscholar.org/24e2/10f8a939acee28f817d07bd031a5b5025738.pdf. Accessed 24 Oct 2021
10. Maga, D., Hiebel, M., Thonemann, N.: Life cycle assessment of recycling options for polylactic acid. Resour. Conserv. Recycl. **149**, 86–96 (2019). https://doi.org/10.1016/j.resconrec.2019.05.018
11. Shrivastava, S., Rawat, J., Agrawal., A.: Controlling DC Motor using Microcontroller (PIC16F72) with PWM. undefined (2012). https://www.semanticscholar.org/paper/Controlling-DC-Motor-using-Microcontroller-with-PWM-Shrivastava-Rawat/12b0e8e4a88560e8f78e8bba3dfe33b1369c8875. Accessed 25 Oct 2021
12. An Algorithm for the In-Field Calibration of a MEMS IMU. https://ieeexplore.ieee.org/abstract/document/8031966/. Accessed 25 Oct 2021
13. Choo, S.H., Amin, S.H.M., Fisal, N., Yeong, C.F., Bakar, J.: Using Bluetooth transceivers in mobile robot, p. 476 (2002). https://doi.org/10.1109/SCORED.2002.1033160
14. Design and implementation of adaptive PID controller for speed control of DC motor. https://ieeexplore.ieee.org/abstract/document/8168478/. Accessed 25 Oct 2021

Development of Two-Wheeled Self-stabilizing Mobile Platform

Dawid Szatkowski and Dominik Rybarczyk(✉)

Institute of Mechanical Technology, Poznan University of Technology, Skłodowska-Curie Square 5, 60-965 Poznań, Poland
dominik.rybarczyk@put.poznan.pl

Abstract. The article describes development of a two-wheeled self-stabilizing platform which is a laboratory model of the human transport vehicle. The device operates with use of the reverse pendulum stabilization principle. The authors presented a change in its behavior along with lifting center of gravity and after mounting on the pendulum top hanging mass. For vertical stabilization, Author used a measuring system consisting of an encoders, a gyroscope and an accelerometer, to calculate actual angular position. System worked under P and PD controllers. The robot has the possibility for: balancing in point, move forwards, backwards and turn. The communication was realized with used of the smartphone dedicated application and Bluetooth interface.

Keywords: Inverted pendulum · Mobile platform · Complementary filter · Selfstabilizing robot

1 Introduction

1.1 Literature Review

Control of a two-wheeled vehicle which kinematic are similar to the inverted pendulum is much more difficult than statically stable vehicles. The control system need to take into account following parameters: influence of center of gravity and load changing, possibility to quiqly changing direction and velocity of movement, possibility to stay in a stable position.

Authors of [1] propose to use of the fuzzy controller. Unfortunately, this is a complicated procedure that requires the adequate number of fuzzy sets, resulting in a significant amount of memory consumption. The more conventional way of regulation is to use a classical PID controller [2] whose characteristic are determined by matching the appropriate coefficient that give a specific response to the value speed of change and time of deviation from equilibrium.

To allow driving with a change of direction, a more advanced PID controller, cascaded, must be used [3]. The similar type of regulation was applied in the case described by the authors: At the input of the controller P is given the set speed from smartphone and current speed of the vehicle. The output is corrected the set tilt angle of the robot,

B. Gapiński et al. (Eds.): MANUFACTURING 2022, LNME, pp. 78–92, 2022.
https://doi.org/10.1007/978-3-031-00805-4_7

which hits the input of the PD controller and is compared to the actual deviation from the vertical. A PWM signal is sent to the output and controls the DC motors. In addition, there is a third controller (P) which maintains the same speed of both wheels. The key element in designing the self-stabilizing device is the appropriate selection of regulator settings. The current position of the device is read from the signal received from the gyroscope and the accelerometer.

Article [4] presents evolutionary approaches for designing inverted pendulum (RIP) controller including genetic algorithms (GA), particle swarm optimization (PSO), and ant colony optimization (ACO) methods.

Article [5] described stabilization problem from the biomechanical point of view. Authors of [6] detailed described is a human-sized dynamically stable mobile robot that balances on a single ball. They performed several tests, that demonstrate the ballbot balancing and locomotion capabilities. The paper illustrates also some interesting human–robot physical interaction behaviors that can be achieved as a result of the ballbot's dynamic stability.

Article [7] presented control of Bicyrobo, which is an unstable system associated with many sources of uncertainties due to un-model dynamics, parameter variations, and external disturbances. Authors used robust and optimal control technique. The authors of [8] presented an unusual application of the problem presented in the article. They described inverted pendulum by placing the pendulum on top of a quadrotor aerial vehicle. The issues related to energy consumption in the operation of the inverted pendulum are discussed in the paper [9]. In publication the dynamic model of a wheeled inverted pendulum is analyzed from a controllability and feedback linearizability point of view [10]. This allowed to control the position and speed of the system.

The issue of the use of the inverted pendulum mechanism for applications in education is presented in the publication [11]. The device described by those citing the said work may have similar applications. For tests of this type of devices, vision systems can also be used, as shown in [12, 13].

On the basis of the literature, it can be concluded that the inverted pendulum as a mechanism is an interesting object from the control point of view. The authors decided to build a device with a modified structure, which would allow to study the influence of an additional moving, hanging mass placed in the upper part.

2 Mobile Platform Structure

2.1 Mechanical Parts

The aim of the publication was to build a mobile platform in an inverted pendulum arrangement, which would be equipped with a hanging mass constituting an additional disturbance. Devices built in this way can be used in the future to move elements at high heights in warehouses in the conditions of limited maneuvering space. The functional purpose of the proposed mobile platform is to compensate for the influence of moving masses on the various heights.

The built-in robot consists of the rectangular plates that form of four levels (Fig. 1). To the bottom plate (first level) two DC motors integrated with the gear are attached. On the second level there are parts of the control system: motor driver modules and the main PCB contain of the microcontroller and inertial sensors. The third level was designed for power supply: There is a battery, power switch, two voltage converters and two fuses. To the fourth plate based on the thin-walled angle bars with dimensions 30 x 30 mm, which allows realization testing of the device. The total height of the model is 129 cm and the weight is equal to 1.5 kg.

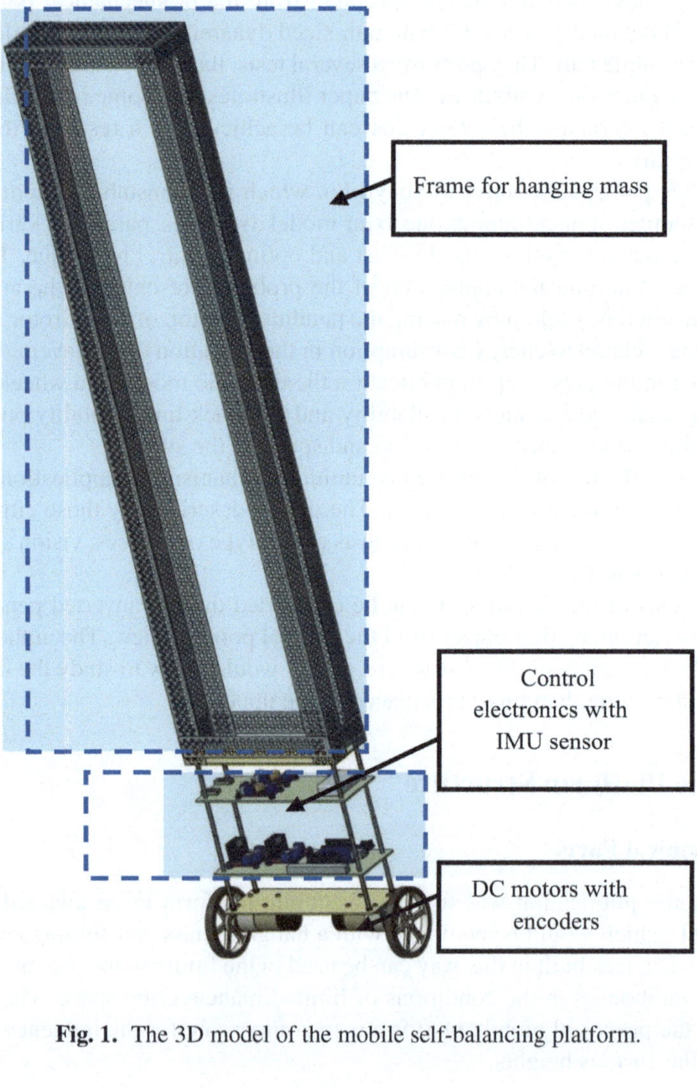

Fig. 1. The 3D model of the mobile self-balancing platform.

2.2 Electronic Controller

Genuino 101 is used as the main module that manages the entire electronics of the built robot (Fig. 2). It is equipped with a 32-bit dual-core Intel microcontroller based on ARC architecture, clocked at 32 MHz. As the IMU sensors Authors used 6-axis accelerometer and gyroscope. The DC motors are controlled by L298 bridges. The system is powered by a lithium-polymer battery with nominal voltage 11.1 V. The DC motors are supplied with 9 V voltage. This treatment reduced the current drawn and allowed stabilizing the system dynamics during the battery discharging.

The control system are equipped with Bluetooth module HC-06 to communicate with mobile devices.

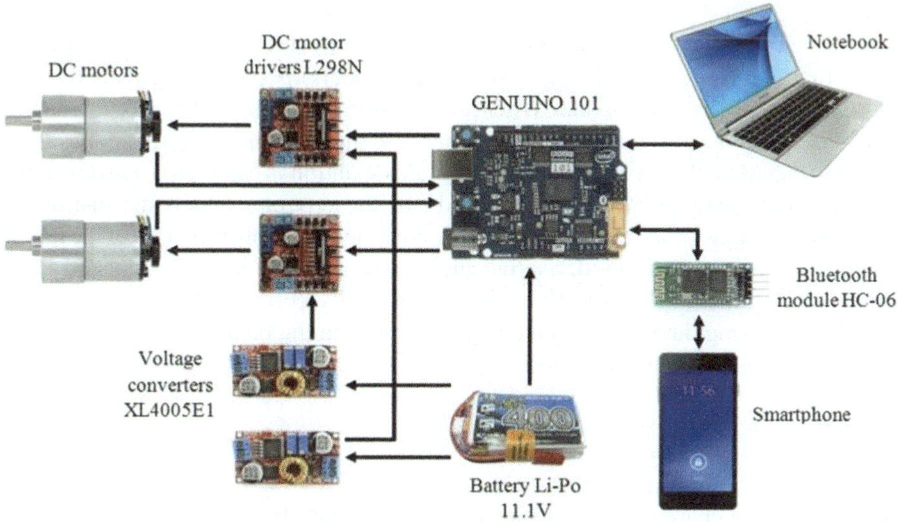

Fig. 2. The electronic controller scheme.

The current data e.g. angle of inclination, wheel speed, set speed, are sent through UART interface with a maximum available speed of 250000 baud/s, to minimize the slowdown of the working program. These data are updated with a frequency of 10 Hz.

2.3 Control Algorithm

The robot's built-in software allows two-wheel balancing and driving at a set speed with the turn option. To make this possible, it is necessary to read and process the data from the sensors and the appropriately calculated PWM signal sent to the motors. In addition, for security purposes, to smartphone the data is sent about the battery powering device charge status.

The structure of the program is based on the multitasking architecture and its divided into five subprograms. In order to no microcontroller load, each thread, depending on the priority level, is realized with an different frequency.

The main thread, with the highest priority, is realized at a frequency equal to 100 Hz. This is the most important and most comprehensive part of the software. Initially, signals from the accelerometer and the gyroscope are read and appropriately converted according to a predefined measuring range. Next, on the base of accelerometer signal, current angle is calculated. For the reduction of noise, the average of the last four readings is calculated. The next step, the pretreated data is modified by the complementary filter formula - the final angle value is derived from the gyro and accelerometer fusion. The complementary filter [5, 14] can be described by the following formula:

$$\theta_f(k) = a[\theta_f(k-1) + \omega_g(k) \cdot dt] + [1-a][\theta_a(k)] \tag{1}$$

where: $\theta_f(k)$, $\theta_f(k-1)$ - output angles; a – complementary filter coefficient; $\omega_g(k)$ - angle velocity from gyroscope; $\theta_a(k)$ – angle from accelerometer; dt – signal sampling period.

For the entire procedure of obtaining the current angle described above, the interrupt handler function (used to read signals from the encoders) is disabled. This is the most important part of the entire code. When the interrupt handler function is re-enabled, the set speed and the current speed are updated and the proportional speed controller is induced to calculate the set swing. Then the angle deviation is calculated and depending on its size and character (forward/backward tilt) the program selects the instruction, introducing the appropriate correction: determining the direction of rotation of the motors and inducing the proportional-derivative angle controller. Before the calculated PWM signal is sent to the motor controllers, it is additionally corrected by the proportional controller of driving direction which compare speeds on the both wheels.

Other threads ale responsible for:

- calculating current speed,
- receiving values from the smartphone,
- sending data to the computer,
- sending battery charge value via Bluetooth to the smartphone.

3 Research Methodology and Experimental Investigation

The physical properties of the object like moment of inertia and kind of used drive are influenced on object behavior. The main aim of the research was to check the impact of changing the center of gravity. In described here case, these settings were dependent on: the diameter of the wheels, DC motors dynamics (velocity, torque), the mass of the entire system, center of gravity location. The controllers settings were taken experimentally and have not been changed during investigations. The individual controller coefficients were equal to:

- proportional velocity controller: kp = 0.072,
- PD angle controller: kp = 35,5; kd = 3,5,
- proportional controller of the driving direction: kp = 17.

As an additional element, the wooden panel with a weight of about 200 g was used, which gave the total weight of the whole set of about 1.7 kg.

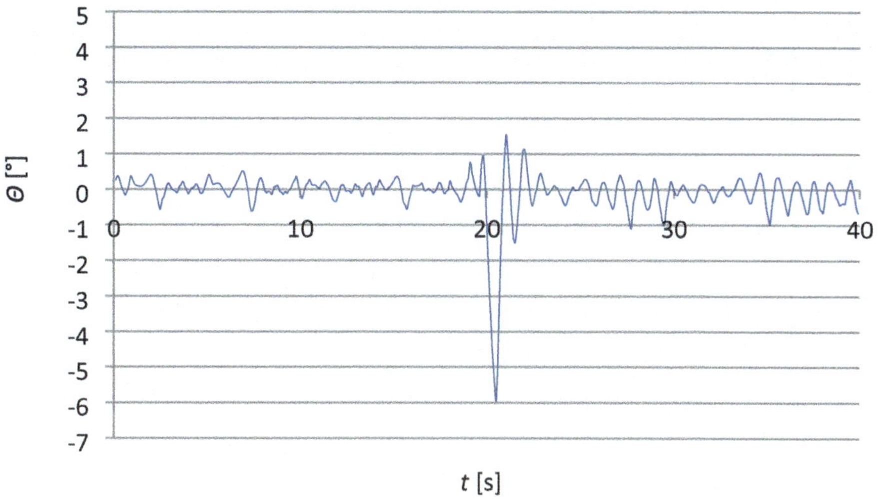

Fig. 3. Stabilization with a static load assembled on height 40 cm

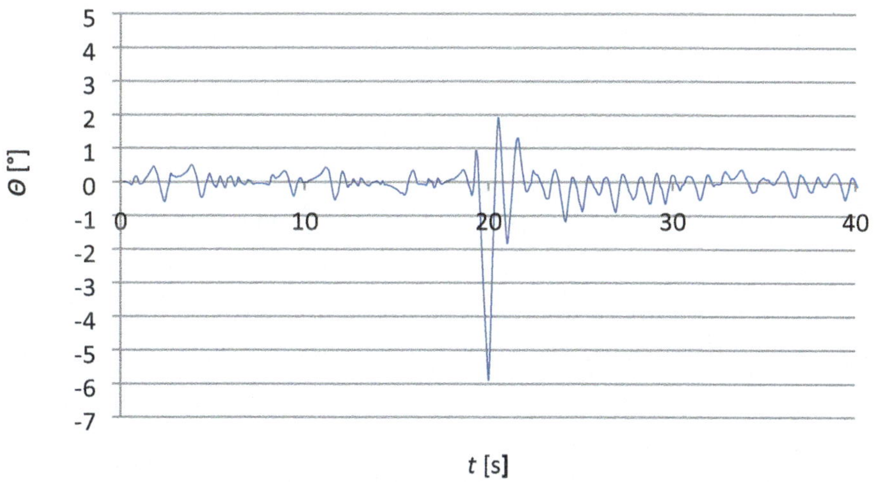

Fig. 4. Stabilization with a static load assembled on height 52.5 cm

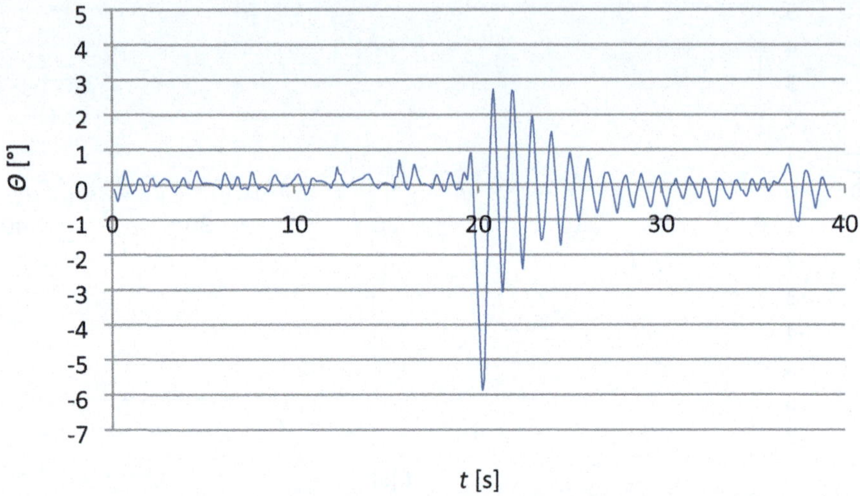

Fig. 5. Stabilization with a static load assembled on height 65 cm

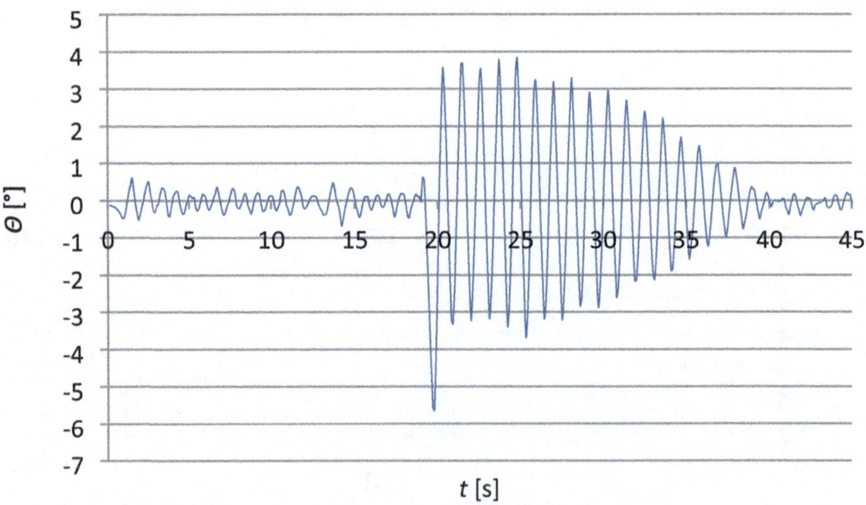

Fig. 6. Stabilization with a static load assembled on height 77.5 cm

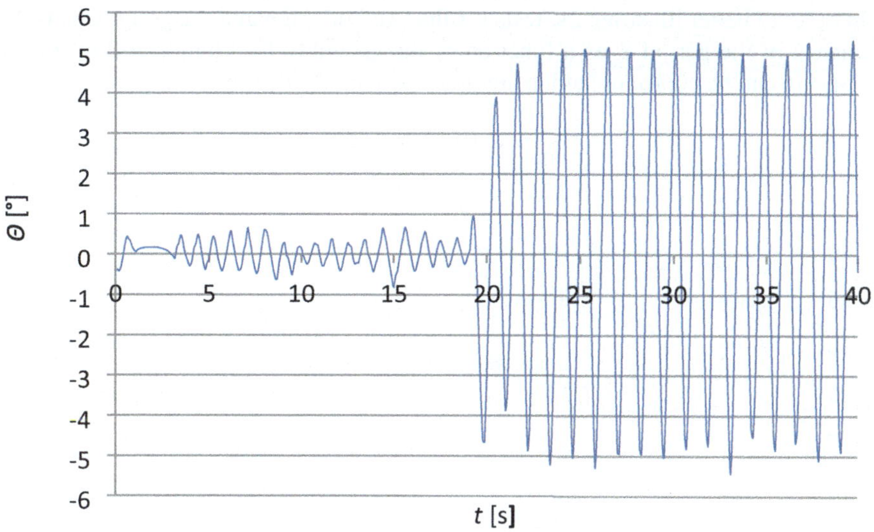

Fig. 7. Stabilization with a static load assembled on height 90 cm

The above graphs (Fig. 3, 4, 5, 6 and 7) have shown the robot's balance with additional mass. During the initial stage, the readings of the deviation angle were recorded during free balancing. The first four graphs show that the robot is within the range of −0.5 to 0.5°, while in the case of placing the mass at 90 cm from the ground, the angle of deviation from the vertical regularly exceeds this value, remaining in the range of ±0.7°.

After about twenty seconds (second stage), the external force was applied to the construction, resulting in a dynamic deflection of less than 6° from the home position (about 0°), thus resulting in a steady-state knock out. At that point, the time measuring of its return to a relatively stable state began. The return period to the steady state started at the time of the peak deviation from the vertical and ended when the deviation point was not over ±1°. On the graphs can be seen that the period of system stabilization increases with the lifting of additional mass. With the mass hanging height equal to 90 cm, the device was unable to return to a stable state. Collected data are shown in the Table 1.

Table 1. Height of the assembled load combined with stabilization period of time

Height of the assembled load h [cm]	Stabilization period of time T [s]
40.0	1.9
52.5	2.1
65.0	4.8
77.5	17.6
90.0	∞

In order to better illustrate the tests results, Authors prepared the graph which show the change of the period *T* as a function of the assembly load height *h* (Fig. 8). This relationship can be approximated using the exponential function:

$$T = 0,0112e0,1062h \quad (2)$$

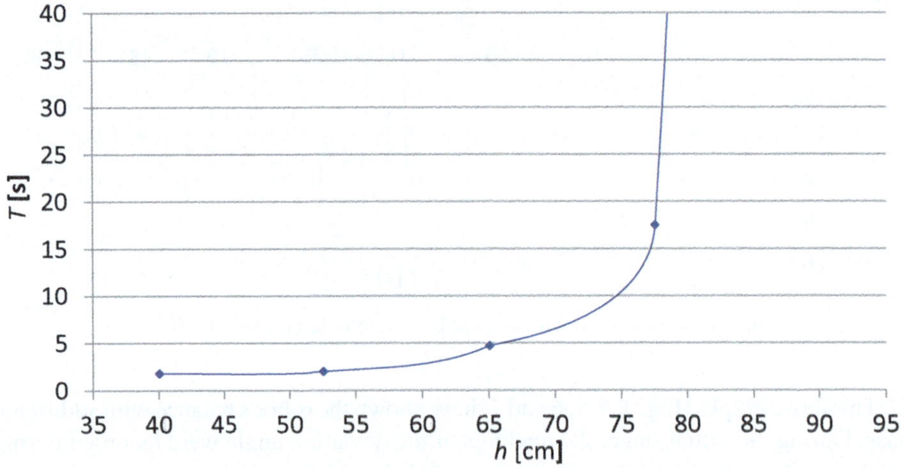

Fig. 8. Stabilization time depending on mass height assembly

The curve on Fig. 9 can be characterized by the logarithmic function. The stabilization period of the system increased in relation to the change in altitude of additional load. At a height of 90 cm, the robot has too high moment of inertia (to all structure), which disqualifies the possibility to return to steady state.

To complete tests, authors examined the behavior of the system after mounting the load at a height equal to 115 cm. The results was show on Fig. 9.

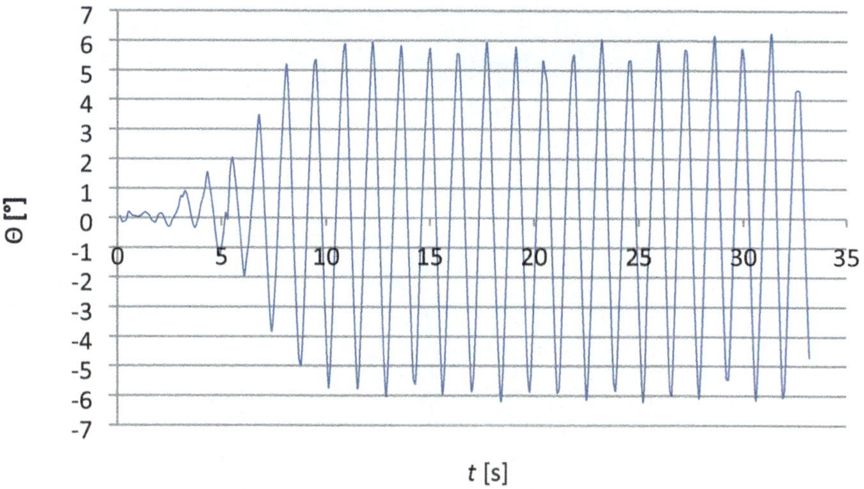

Fig. 9. Stabilization with a static load assembled on height 115 cm

After four seconds, robot lost his stability and his vertical deviation increased to over 6o. In this case the allowable mounting height of the additional mass has been exceeded and therefore system cannot balancing correctly.

The presented here robot can be characterized and described from the kinematic point of view as a type of the pendulum. During movement, the center of gravity is constant. However, a behavior of the system has been changed, when another physical pendulum is attached to it (Fig. 10). During the swing of the additional mass, the center of gravity of the robot changes dynamically. The object became highly nonlinear.

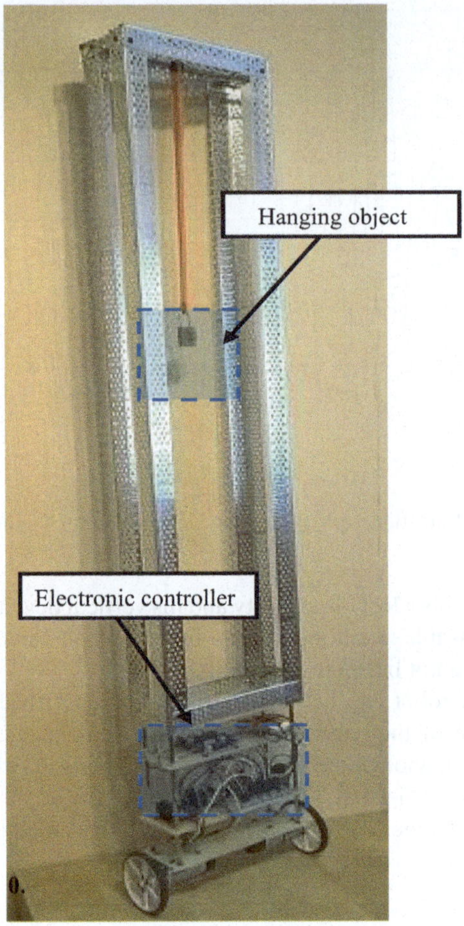

Fig. 10. Mobile platform with assembled pendulum

The additional mass (load) was connected to robot frame by used of plastic rod 400 mm length). Authors checked following values of additional mass: 10, 20 and 50 g. The following graphs show system behavior (Fig. 11, 12 and 13).

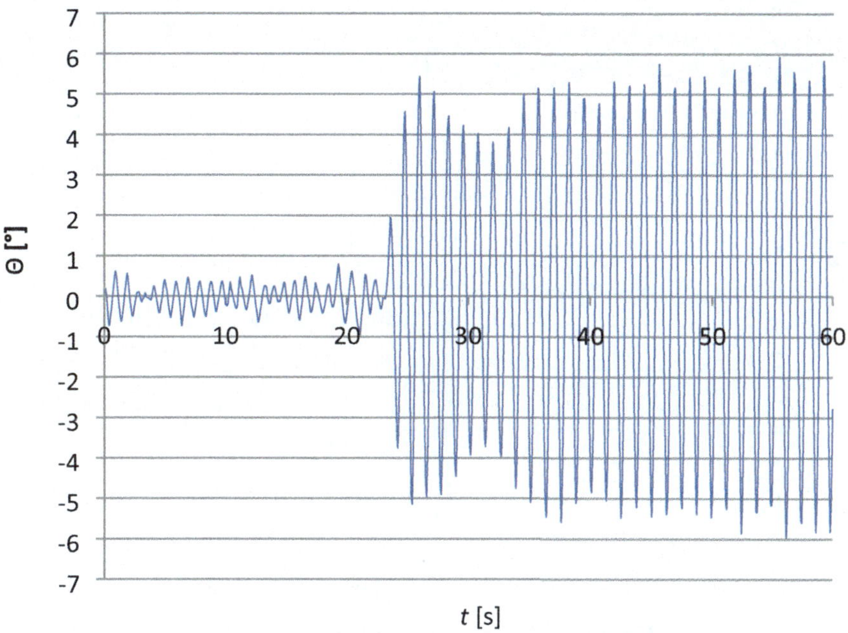

Fig. 11. Stabilization with hanging load - 10 g

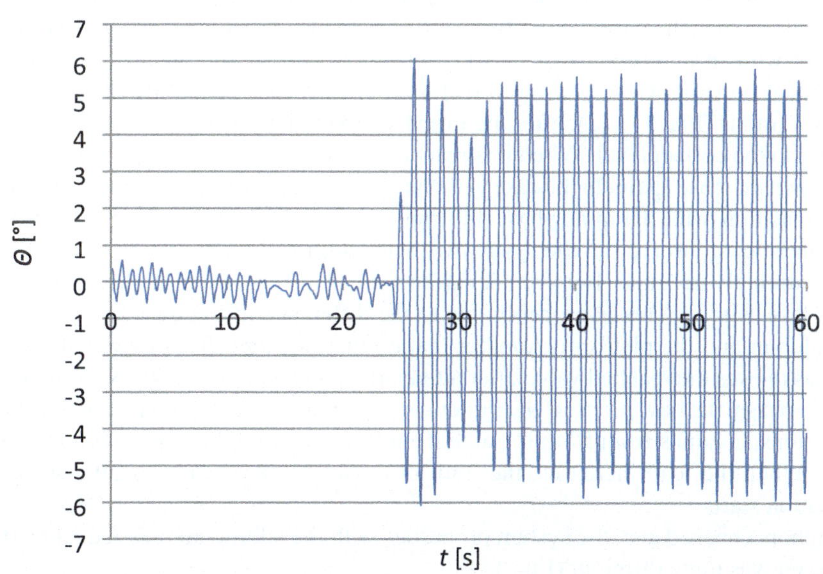

Fig. 12. Stabilization with hanging load - 20 g

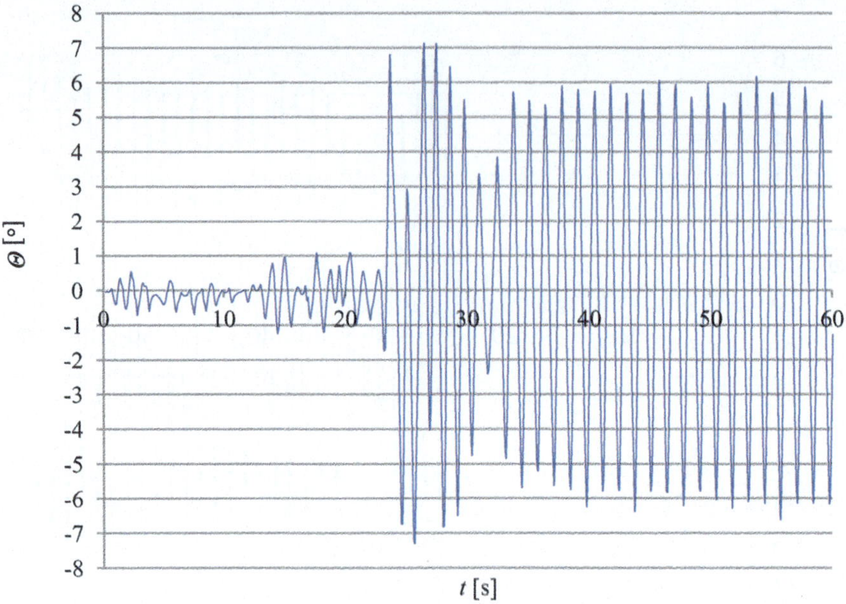

Fig. 13. Stabilization with hanging load - 50 g

In the first phase of the graph, for about 25 s, free balancing is shown (without external distortion). It can be seen that the device, despite of the small movements of attached pendulum, is able to maintain a stable position without major problems. The second phase, which started about twenty-five seconds, Authors extortion pendulum to swinging with an angle equal to 45°. All graphs show sudden, significant deviations from the vertical position. However, the waveforms in each case is slightly different: in the first two graphs (Fig. 11, 12), after the initial sharp deviation, the oscillations regularly increase to the certain level. When it is reached, the oscillations are suppressed to about 4° and then increase again. In the case with the attached mass of 20 g weight, can be seen a faster transition to the damping phase (the time of initial increasing is shorter). A different characteristic is observed in the system with an additional mass of 50 g (Fig. 13). The run is extremely irregular. Between deviations for 7° appear much smaller swings, with values less than 3°. This is probably due to the difference in frequencies of deviations in pendulum and the platform. Another feature that distinguishes this waveform from previous ones is the reached value after attenuation, which is only 2.4°. Nevertheless, in the next phase, the deviations of the system are again increasing, which, like in previous cases, stops at values from 5° to 6°, with a tendency of slight increase (with increasing mass). When the load increases, the limit of deviation also increases after which the attenuation starts.

Authors checked also the system properties with the additional mass equal to 100 g. The result was quite different (Fig. 14).

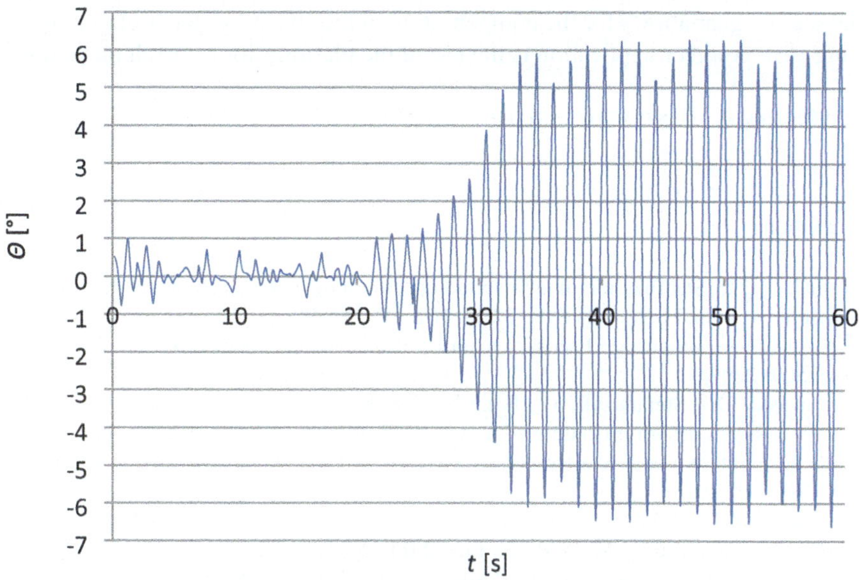

Fig. 14. Stabilization with hanging load - 100 g

After free, undisturbed balancing, lasting about twenty seconds, the swings of the system exceeded over 6° (Fig. 15). The loading of physical pendulum by mass of 100 g makes it impossible to keep the device in a relatively stable position, because the slightest swing causes significant shifts in center of gravity. In addition, after manually stabilizing the object, an attempt was made to tilt the pendulum to the angle of equal to 30°. At this value the center of gravity of the system changed dynamically and the robot was unable to stay on the wheels (resulting - immediate fall).

4 Conclusion

This article describes the stabilization of two-wheeled mobile platform with the variable center of gravity and swinging mass hanging to the platform frame.

The first step of tests was to increase the height of the additional mass. It has been shown that, when the center of gravity of the model is raised, its stabilization time increases logarithmically. For described here system the maximum value of the load hanging is 90 cm height. After exceeding this value, the system lose its equilibrium point and is no able to return to it.

The second step was to examine the behavior of the system after mounting an additional pendulum. The graphs show that even with the small mass of the pendulum, imbalance results in a lack of stability.

The built-in mobile platform can be used in two ways. The device can be supplemented with gripping elements, thanks to which it will be possible to move elements at high heights in spaces with limited maneuvering space, e.g. in warehouses. The structure of the steering gear allows for the compensation of the changing center of gravity of the steering gear. Another application is the use of the platform for research purposes.

Acknowledgment. This research was funded by grant number 0614/SBAD/1547.

References

1. Yamakawa, T.: Stabilization of an inverted pendulum by a high-speed fuzzy logic controller hardware system. Fuzzy Sets Syst. **32**(2), 161–180 (1989)
2. Kędzierski, J., Tchoń, K.: Feedback control of a balancing robot IFAC Proc. Vol. **42**(13), 495–500 (2009)
3. Seredyński, D. , Stefańczyk, M., Banachowicz, K., Świstak, B., Kutia, V., Winiarski, T.: Control system design procedure of a mobile robot with various modes of locomotion. In: 21th IEEE International Conference on Methods and Models in Automation and Robotics, pp. 490–495 (2016)
4. Hassanzadeh, I., Mobayen, S.: Controller design for rotary inverted pendulum system using evolutionary algorithms. Math. Probl. Eng. **2011** (2011)
5. Park, S., Horak, F.B., Kuo, A.D.: Postural feedback responses scale with biomechanical constraints in human standing. Exp. Brain Res. **154**(4), 417–427 (2004)
6. Nagarajan, U., Kantor, G., Hollis, R.: The ballbot: an omnidirectional balancing mobile robot. Int. J. Robot. Res. (2013)
7. Thanh, B.T., Parnichkun, M.: Balancing control of bicyrobo by particle swarm optimization-based structure-specified mixed h2/h∞ control. Int. J. Adv. Robot. Syst. (2008)
8. Hehn, M., D'Andrea, R.: A flying inverted pendulum. In: 2011 IEEE International Conference on Robotics and Automation (2011). https://doi.org/10.1109/ICRA.2011.5980244
9. Kuo, A.D., Donelan, J.M., Ruina, A.: And3 energetic consequences of walking like an inverted pendulum: step-to-step transitions. Exerc. Sport Sci. Rev. **33**(2), 88–97 (2005)
10. Pathak, K., Franch, J., Agrawal, S.K.: Velocity and position control of a wheeled inverted pendulum by partial feedback linearization. IEEE Trans. Rob. **21**(3), 505–513 (2005). https://doi.org/10.1109/TRO.2004.840905
11. Furuta, K., Yamakita, M., Kobayashi, S., Nishimura, M.: A new inverted pendulum apparatus for education. In: Kheir, N.A., Franklin, G.F., Rabins, M.J. (eds.) Advances in Control Education 199, pp. 133–1381. Amsterdam, Pergamon (1992). https://doi.org/10.1016/B978-0-08-040958-0.50029-4
12. Magana, M.E., Holzapfel, F.: Fuzzy-logic control of an inverted pendulum with vision feedback. IEEE Trans. Educ. **41**(2), 165–170 (1998). https://doi.org/10.1109/13.669727
13. Ye, W., Li, Z., Yang, C., Sun, J., Su, C.-Y., Lu, R.: Vision-based human tracking control of a wheeled inverted pendulum robot. IEEE Trans. Cybern. **46**(11), 2423–2434 (2016). https://doi.org/10.1109/TCYB.2015.2478154.Dd
14. Tseng, S.P., Li, W., Sheng, C., Hsu, J., Chen, C.: Motion and attitude estimation using inertial measurements with complementary filter. In: 2011 8th Asian Control Conference (ASCC), pp. 863–868 (2011)
15. Wu, J., Zhou, Z., Chen, J., Fourati, H., Li, R.: Fast complementary filter for attitude estimation using low-cost MARG sensors. IEEE Sens. J. **16** (18), 6997–7007 (2016). https://doi.org/10.1109/JSEN.2016.2589660

Multicriteria Optimization of the Worm Gearing Drive for Tool Magazine

Oleg Krol(✉), Volodymyr Sokolov, and Aleksander Golubenko

Volodymyr Dahl East Ukrainian National University, 59-a Tsentralnyi Prospect, Severodonetsk 93400, Ukraine

krolos@snu.edu.ua

Abstract. A search procedure for a rational design of a worm gear of a drive mechanism for a tool magazine using the main criteria for the effective functioning of a metal-cutting machine in the mode of automatic tool change is considered. The analysis of the chain tool magazines design of the drilling, milling and boring machining center based on the constructed solid models in the environment of the integrated CAD KOMPAS-3D is carried out. 3D modeling of a worm gear using specialized software for the design of mechanical transmissions integrated into the contour of the KOMPAS system is carried out. A procedure for multicriteria optimization of the design parameters of the worm gear drive for the machining center by the method of Lagrange multipliers is proposed. An algorithm for finding the best design solutions for a worm gear, a feature of which is the introduction of two derived criteria expressed as a ratio of the optimized variables is proposed. A numerical analysis of the results of the system equations solving reflecting the dependence of the main gear transmission criteria on the optimized parameters is carried out. Based on the results obtained, it is shown in what ranges of design parameters of the worm gear elements and their gear ratios the effective functioning of the tool magazine is possible within the permissible values of the basic parameters of the worm gear.

Keywords: Chain tool magazine · Worm gearing · Structural optimization

1 Introduction

Modern metal-cutting machine tool systems represent complex assembly structures designed for shaping various types of parts with the ability to automatically change tools. To implement the change procedure, tool magazines, auto-operators, auxiliary and cutting tools are used, which together form the machine Tool System (TS). In the practice of mechanical engineering, a classification of tooling systems is used, in particular for drilling, milling and a boring cluster of CNC machine tools and various designs of Tool Magazines (TM) [1–3]. For the machine housings processing of medium and high complexity, a significant number of various tools are required (taking into account duplicating tools) In the TM housing, such tool blocks are mounted that provide technological operations for milling planes and flattened surfaces, drilling and countersinking

B. Gapiński et al. (Eds.): MANUFACTURING 2022, LNME, pp. 93–102, 2022.
https://doi.org/10.1007/978-3-031-00805-4_8

holes, machining grooves and other machined planes. This gives the TS technological flexibility, on the other hand, requires the use of high-capacity TM designs. For such purposes, the most rational is a chain tool magazine.

In the process of designing machining centers, it is necessary to be guided by a certain strategy for the development of tool systems. In work [1] outlines are contours and a study of the optimal strategy of tool support for machine-building production is carried out. For this purpose, based on a systematic approach, a number of factors have been identified that influence the choice of a stable production process in the direction of two aspects. A feature of this task is the presence of two conflicting goals:

1) adherence to the schedule of production orders and, as a result, an increase in the level of inventories and transportation costs; In this regard, interesting work [2] in which the author created a mathematical model reflecting the minimum stock of the tool and the value of the average delay associated with the identification of the tool. This model most adequately reflects the situation associated with the storage of tools in a centralized tool workshop or localized storage.
2) High tool availability and reduced machining times.
   The resolution of this conflict of goals lies, according to the authors [1], in determining the optimal storage area within a set of options: a) centralized tool shop; b) localized storage, and c) machine tool magazine.

In the process of functioning a CNC machine (machining center type) with a variety of tooling, several errors arise. In this case, the errors of Machining Centers (MC) are formed more intensively than in conventional CNC machines and can lead, in aggregate, to relative linear and angular displacements of the working machine parts, carrying the cutting tool and the workpiece. Correction of the angular displacements is especially difficult.

The efficiency and productivity of the machining center are largely related to the interrupted operation of the automatic tool changer (ATC), and the accuracy and quality of forming blanks with the level of errors arising from the interaction of "tool magazine – auto-operator – machine tool spindle.

Some works to the analysis of errors in the functioning of the TS machining centers are related.

So, works [3, 4] are devoted to the analysis of errors associated with the positioning of tools in the TM, which can lead to failures in the ATC operation of the machining center. In this regard, it is important to have an early warning system for tool magazine failures. Such a system for analyzing vibration signals in situations of removing tools from the magazine slot and mounting it in the conjugated conical bore of the spindle is presented in [3]. The same problem of predicting TM for machining center failures based on performance assessment and identification of the most critical components using a fuzzy comprehensive assessment was proposed in [4]. It is shown that among the various causes of failure, the component associated with tool jamming with inaccurate positioning accounts for 21% of the total volume of failures. The software VERICUT [5] for 5-axis milling machines of medium sizes equipped with chain tool magazines is developed. This software is designed to simulate the main shaping processes, including to ensure high position accuracy in the chain TM during tool change.

In the work [6] the issues of increasing accuracy and reducing the cost of processing due to improvement of the components of the technological system of the CNC lathe are considered. Analytical dependencies to determine the errors of the machining process in the case of using a multi-cutter turret are formed. This allows reducing the machining errors along the X- and Y-axes by up to 30%.

In turn, the positioning indicators – positioning accuracy, positioning resolution, positioning repeatability depend on the design characteristics of the TM drive, the main mechanical transmission of which is a worm gear. In the reversing mechanisms of machine tools and machine-tool complexes, the use of classical mechanical transmissions, such as a worm gear, is difficult due to the problem of strict synchronization of the drive and drive shafts rotations [7, 8]. This situation occurs during the reverse movement of the worm when the driven worm wheel is stationary for a certain period. The misalignment of the rotations of the worm gear elements causes inaccuracy in the TM positioning.

Analyzing the implementation methods for increasing the accuracy of the actuating movements of the tool system by reducing the gaps in the transmission, two approaches can be distinguished. The first is associated with the search for a new technology for the production of worm gear elements. The second is associated with the use of optimal design solutions in worm gearing. In this regard, the designer is faced with the need to analyze the engagement zone, which is characteristic of various types of mechanical gears (spur, helical, bevel, worm, etc.) [9, 10]. Large-scale studies of various types of mechanical transmissions structures in advanced computer-aided design systems and specialized applied software are carried out. Thus, dual-part gearing research in the ABAQUS program environment is realized. Such research is aimed at modeling and determining contact stresses on the lateral surfaces of the teeth and bending stresses at the root of the teeth [11]. Of interest are also two author's specialized subroutines [12] as part of the ABAQUS program. The proposed subroutines make it possible to assess the level of contact stresses, pressure on the lateral surfaces of the tooth, the length of the contact lines and the wear resistance of the contacting surfaces.

In [13], the fatigue strength of the output link of a two-stage gear mechanism under contact and bending stresses was researched. An experiment to determine the fatigue strength in the most dangerous section at the tooth tip, compared with the permissible values with a safety factor equal to value 1.4 was carried out. The authors use the procedure for optimizing the gear mechanism design. At the same time, the scenario of the optimization procedure is not presented in the work.

The optimization of the worm gears parameters in works [14–16] in various aspects is considered. So in works [14, 15] analytical and experimental approaches to solving the problem of choosing a rational type of gearing and optimization of its parameters are presented. The work [16, 17] presented the results of multicriteria synthesis of gears with different types of worms, as well as the influence on these criteria of individual geometrical-kinematic characteristics of the engagement, such as the length and shape of contact lines, reduced curvature, profile angle of turns, etc.

Based on the analysis of the above results of the machine tool with TM, it becomes necessary to research the properties of the tool magazine structure, which is equipped with a large number of various kinds of tool blocks. On the other hand, there is a need to search for rational design solutions when creating a TM drive based on the worm gear.

Problem statement: to create a complex of the 3D-models main components of a chain tool magazine, as a basis for determining its technical characteristics and developing an optimization procedure for choosing the rational TM drive mechanism design.

2 Three Dimensional Modeling of the Tool Magazine

Consider the process of manufacturing a complex housing part on a drilling, milling and boring machining center. The technological process of machining involves the use of a large number of cutting and auxiliary tools. For the considered nomenclature of the tool, there is the possibility of reuse with the potential restoration of cutting properties (regrinding, changing the cutting insert, adjusting to size). The presence of such factors, according to the strategy of optimal tooling support [18, 19], makes it possible to choose a tool magazine as the optimal storage area.

When conducting engineering research and evaluating the efficiency of the functioning of machining centers, it becomes necessary to create new sets of tool magazines 3D-models of disk and chain types for machines of 2- and 3-standard sizes [20, 21]. For such magazines, you need to create separate sections of 3D models of auxiliary and cutting tools using rendering technology. As noted above, the sufficient complexity of the technological process and a significant number of cutting tools justify the choice of a chain tool magazine.

Consider the design of a machining center chain magazine. Tool magazine is intended for storage and delivery to the position of changing tool holders with cutting tools. The TM of the chain type of the machining center is mounted on a column, taken out from the cutting zone of the machine. The tool magazine's chain in two versions consists of 36 (60) elements and is driven by a gearbox.

In this work, as a research part of the TS operability and reliability, a complex of 3D models of tool magazines, equipped with auxiliary and cutting tools for milling and axial machining operations (drilling, countersinking, etc.) is created.

The use of 3D modeling with the principle of associativity in this task allows you to generate TM-constructions, any necessary views and sections in the drawings, to obtain initial data for the tasks of design and technological preparation of production [22–24]. Based on the constructed model of the structure, precise lists of equipment, products and materials used in this model are formed – specifications, bills of materials. As a computer-aided design system for the implementation of three-dimensional TS-modeling, it is advisable to use the proven efficiency integrated CAD KOMPAS-3D with the technology of integrated end-to-end 3D design, developed by the ASCON group of companies [25, 26] and the Artisan-rendering module built into it.

In the environment of the integrated CAD KOMPAS-3D, a 3D model of a chain tool magazine for 36 tool positions has been developed (Fig. 1).

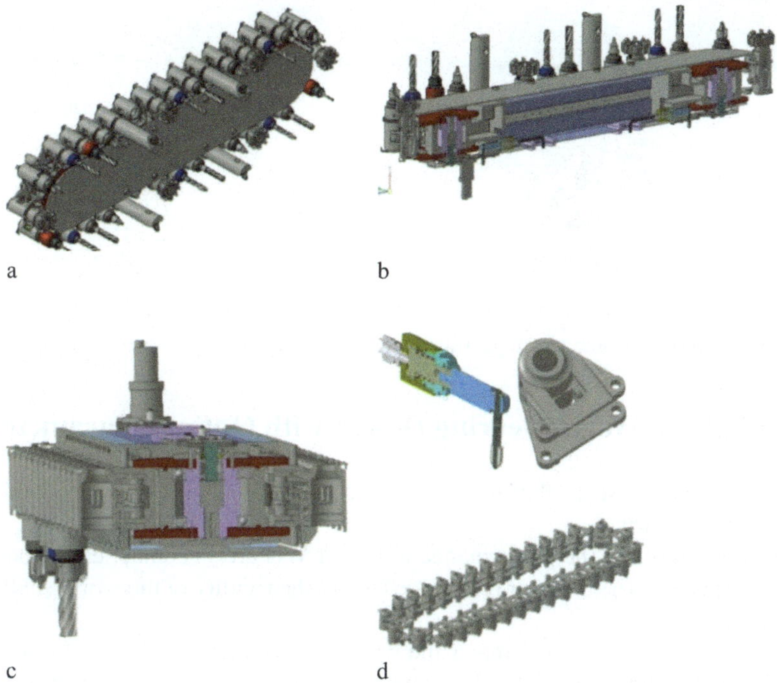

Fig. 1. Three-dimensional model of a chain tool magazine: a – general view; b – longitudinal section; c – cross-section; d – chain and tool holder slot

3 Solid Modeling of the Tool Magazine's Drive

To remove the tool or return it to the corresponding slot, the chain must be mounted in the "Change" zone. The movement of the chain from an electric motor through a worm gear is carried out.

The gearbox consists of a housing in which a worm gear transmits the movement from the electric motor to the driving sprocket of the TM chain and a cylindrical gear leading to the circular displacement transducer.

In the factory version of the worm gear, a worm with an uneven pitch is used, which allows, as wear and tear, to adjust the backlash in the Worm Gearing (WG) by engaging the turns of the worm with a greater thickness. The adjustment by fitting the ring when loosening the screws in the coupling is carried out. At the same time, such a constructive option is associated with additional costs and an increase in auxiliary and preparatory-final time for processing housing parts [27, 28].

To assess the WG-design parameters, 3D models in the KOMPAS environment have been developed (Fig. 2) [29, 30].

a b

Fig. 2. 3D models of the worm gearing: a – worm drive section; b – worm engagement

4 Search for a Worm Gearing Design with Optimal Parameters

In this paper, we propose a solution to the multicriteria problem of optimizing the worm gearing of the TM drive.

The criteria limiting the performance of power WG are 1) contact endurance (σ_H); 2) gearing efficiency (η); 3) wear resistance (γ); 4) the rigidity of the worm shaft (δ); 5) heat resistance (Δt^0).

When operating a chain tool magazine with a large number of different tools and its changeability dynamics, the criteria of contact endurance σ_H and efficiency η come to the forefront.

This paper presents the results of optimization of the WG basic parameters: $\{z_1, z_2, q\}$: the number of worm threads z_1, the number of worm wheel teeth z_2 and the worm diameter factor q (the quotient of reference diameter to the axial module) which depend not only on the gear ratio u but also on what performance criterion is set as the main one.

We will choose of criteria as optimization criteria: {Contact endurance + Efficiency} $\rightarrow \{\sigma_H + \eta\}$.

The contact endurance σ_H, expressed in terms of one of the variable variables (y), is:

$$\sigma_H = C_\sigma \cdot \sqrt{(1+y)^3/y}, \tag{1}$$

where $C_\sigma = 5300 \cdot \sqrt{\frac{K_H}{a_w^3} \cdot T_2} = const;\ y = \frac{q}{z_2}$;

T_2 – torque on the worm wheel, N m;

a_w – center distance, mm;

K_H – loading coefficient.

Expressing the value T_2 in terms of the torque of the worm T_1 and the gear ratio u: $T_2 = T_1 \cdot u \cdot \eta$, and the efficiency η in terms of another variable $s = q/z_1$:

$$\eta = \sqrt{\frac{s - f'}{s \cdot (1 + f' \cdot s)}} \tag{2}$$

where $f' = tg\varphi'$ – reduced coefficient of friction; φ' – the reduced angle of friction in engagement, deg.

After the transformations, we get the objective function $\sigma_H = \sigma_H(s, y)$ that combines two criteria for the WG performance – the contact endurance of the teeth (σ_H) and the efficiency of the engagement (η):

$$\sigma_H = C_\sigma^* \cdot \sqrt{\frac{(s - f') \cdot (1 + y)^3}{s \cdot y \cdot (1 + f' \cdot s)}} \tag{3}$$

where $C_\sigma^* = 5300 \cdot \sqrt{K_H \cdot T_1 \cdot u / a_W^3} = const.$

As a result, the Lagrange function L takes the form [5]: $L = \sigma_H + \lambda \cdot g$. Here λ – Lagrange multiplier; $g = s - y \cdot u = 0$ – the equation of communication of the varied variables; u – gear ratio.

Using the obtained relations, we find the minimum of the two-parameter function (3), as a solution to the system of three equations with three unknown parameters – $s,\ y,\ \lambda$:

$$\begin{cases} \frac{\partial L}{\partial s} = \frac{\partial \sigma_H}{\partial s} + \lambda \cdot \frac{\partial g}{\partial s} = 0; \\ \frac{\partial L}{\partial y} = \frac{\partial \sigma_H}{\partial y} + \lambda \cdot \frac{\partial g}{\partial y} = 0; \\ \frac{\partial L}{\partial \lambda} = g(s, y) = 0. \end{cases} \tag{4}$$

System of Eqs. (4) in expanded form:

$$\begin{cases} C_\sigma^* \cdot \sqrt{\frac{(1+y)^3}{y}} \cdot \frac{f' \cdot (1 + 2 \cdot f' \cdot s - s^2)}{2 \cdot s \cdot (1 + f' \cdot s) \cdot \sqrt{s \cdot (s - f') \cdot (1 + f' \cdot s)}} + \lambda = 0; \\ C_\sigma^* \cdot \sqrt{\frac{s - f'}{s \cdot (1 + f' \cdot s)}} \cdot \frac{(1+y)^2 \cdot (2 \cdot y - 1)}{2 \cdot \sqrt{y^3 \cdot (1+y)^3}} - \lambda \cdot u = 0; \\ s - y \cdot u = 0. \end{cases} \tag{5}$$

The system of three-parameter Eqs. (5) is reduced to a cubic equation with one variable parameter s:

$$f' \cdot (1 + 2 \cdot f' \cdot s - s^2) \cdot (u + s) + u \cdot (2 \cdot s - u) \cdot (s - f') \cdot (1 + f' \cdot s) = 0. \tag{6}$$

The value of the root of the variable s found from Eq. (6) is used to determine the value of the second variable – y, using the relationship equation: $y = s/u$.

The calculated values of s and y make it possible to find the optimal values of the three parameters of the worm gearing – $\{z_1,\ z_2,\ q\}$. In this case, the boundary conditions for the parameters s and y are:

$q = s \cdot z_1 = [6.3 \ldots 24]$ – q-range from National Standard NS 19672–74;

$z_2 = q/y \approx 28 \ldots 80$ – recommended range of the number of teeth of the wheel in the power WG.

Numerical analysis of the results of solving Eq. (6) showed that only for four-thread worms ($z_1 = 4$) and for very small gear ratios (u), the values of the roots of the variables s and y give the transmission parameters q and z_2, which fit into the indicated constraints.

1. When the reduced coefficient of friction: $f' \approx 0.07$ ($\varphi' \approx 3.8^0$), which corresponds to the sliding speed of the worm $V_S \approx 0.5...0.6\ m/s$ – along with the cast-iron teeth of the wheel. In this case gear ratio $u : u = u_{\max} = 8 \rightarrow q = s \cdot z_1 = 5.8 \cdot 4 = 23.2$, which is close to $q_{\max} = 24$ in NS 19672–74.

2. When the friction forces in the engagement are reduced: $f' \approx 0.02$ ($\varphi' \approx 1^0$), which corresponds to the sliding speed of the worm $V_S \approx 14 \ldots 15\ m/s$ on the tin teeth of the wheel: $u = u_{\max} = 10 \rightarrow q = s \cdot z_1 = 5.7 \cdot 4 = 22.8$, which is also close to $q_{\max} = 24$ in NS 19672–74.

For $u > 10$, the worm diameter factor, q is significantly higher than the largest standard value $q_{\max} = 24$. That is, for $z_1 = 1$ and for $z_1 = 2$ any gear ratios, as well as for $z_1 = 4$, but with $u > 10$, it is practically impossible to optimize WG simultaneously according to two criteria $\{\sigma_H + \eta\}$.

5 Conclusion

1. A procedure for 3D modeling of a chain tool magazine with a capacity of 36 tool blocks for drilling, milling, boring machining centers has been implemented. This 3D project, executed in the environment of the integrated CAD KOMPAS-3D, became the winner of the international competition "Future Aces of Computer 3D Modeling"
2. A three-dimensional modeling of the worm gearing of the chain tool magazine drive using the specialized program "Shafts and mechanical transmission-3D", which provides a sharp reduction in the design time for mechanical gearings of machining center drives was carried out.
3. The multicriteria optimization of the worm gearing of the TM-drive is carried out by the method of Lagrange multipliers. It is shown that with the limiting criteria of contact endurance and gearing efficiency, the optimal gearing parameters $\{z_1, z_2, q\}$ are possible only for four-thread worms at low gear ratios $u \leq 10$. Experimentally, close to optimal values of the worm diameter factor q were obtained in the range of low (up to 0.6 m/s) and high (up to 15 m/s) sliding speeds of the worm, which provide an effective searching procedure for a rational design of the tool magazine drive. Further research is related to the creation of a database section for the designs of auto-operators and the optimization of the process of changing tool blocks, taking into account the presence of duplicate cutting tools.

References

1. Schaupp, E., Abele, E., Metternich, J.: Evolution relevant factors for developing an optimal tool storage strategy. In: 5th CIRP Global Web Conference Research and Innovation for Future Production. Procedia CIRP, vol. 55, pp. 23–28 (2016)
2. Rochow, P., Hund, E., Gruss, M., Nuhuis, P.: Development of a mathematical model for the calculation of the tool appropriation delay depending on the tool inventory. Logist. J.: referierte Veröffentlichungen **2015**, 1–11 (2015). ISSN 1860-7977. (urn:nbn:de:0009-14-41379), https://doi.org/10.2195/lj_Rev_rochow_en_201503_01
3. Chen, C., et al.: Study on failure warning of tool magazine and automatic tool changer. J. Vibroeng. **18**(2), 883–899 (2016)
4. Li, G., et al.: Fault forecasting of a machining center tool magazine based on health assessment. Math. Probl. Eng. 1–10 (2020). https://doi.org/10.1155/2020/5796965

5. Yan, Y., Yin, Y., Xiong, Z., Wu, L.: The simulation and optimization of chain tool magazine automatic tool change process. Adv. Mater. Res. **834–836**, 1758–1761 (2014)
6. Gasanov, M., Kotliar, A., Basova, Y., Ivanova, M., Panamariova, O.: Increasing of lathe equipment efficiency by application of gang-tool holder. In: Gapiński, B., Szostak, M., Ivanov, V. (eds.) MANUFACTURING 2019. LNME, pp. 133–144. Springer, Cham (2019). https://doi.org/10.1007/978-3-030-16943-5_12
7. Levitan, Y., Obmornov, V., Vasiliev, V.: Worm reducers: Handbook. Mechanical Engineering, Leningrad (1985)
8. Chernavsky, S., Itskovich, G., Kiselev, V., et al.: Design of mechanical gears: Textbook. Mechanical Engineering, Moscow (1976)
9. Litvin, F.: The Theory of Gearing. Nauka, Moscow (1968)
10. Litvin, F., Fan, Q., Aznar, F.: Computerized Design, Generation, Simulation of Meshing and Contact of Face-Milled Formate Cut Spiral Bevel Gears. NASA/CR, Santa Monica (2001)
11. Pacana, J., Kozik, B., Budzik, G.: Strength analysis gears in dual path gearing by means of FEM. Diagnostyka **16**(1), 41–46 (2015)
12. Fudali, P., Pacana, J.: Application development for analysis of bevel gears engagement using FEM. Diagnostyka **16**(3), 47–51 (2015)
13. Kačalová, M., Pavlenko, S.: Strength and dynamic analysis of a structural node limiting the multi-output gear mechanism. Acta Polytechnika **57**(5), 316–320 (2017). https://doi.org/10.14311/AP.2017.57.0316
14. Bernatsky, I.P., Vjushkin, N.I., Gerasimov, V.K., Komkov V.N.: Rational choice of the parameters of engagement of worm cylindrical gears. In: Toothed and Worm Gears, pp. 193–210. Mechanical Engineering, Leningrad (1974)
15. Shevchenko, S., Mukhovaty, A., Krol, O.: Gear transmission with conic axoid on parallel axes. In: Radionov, A.A., Kravchenko, O.A., Guzeev, V.I., Rozhdestvenskiy, Y.V. (eds.) ICIE 2019. LNME, pp. 1–10. Springer, Cham (2020). https://doi.org/10.1007/978-3-030-22041-9_1
16. Dusev, I.I., Baturina, N.Y.: Multicriterial synthesis of worm gears with optimal parameters. Novocherkask Polytechnical Institute, Novocherkassk (1987)
17. Krol, O., Sokolov, V.: Research of modified gear drive for multioperational machine with increased load capacity. Diagnostyka **21**(3), 87–93 (2020). https://doi.org/10.29354/diag/126026
18. Dervoort, W.H.: Modern Machine Shop Tools, Their Construction, Operation and Manipulation, Including Both Hand and Machine Tools. Creative Media Partners, London (2018)
19. Krstić, M.: Rational choice theory and addiction behavior. MARKET/TRŽIŠTE **26**(2), 163–178 (2014)
20. Brecher, C., Fey, M., Daniels, M.: Modeling of position-, tool- and workpiece-dependent milling machine dynamics. High Speed Mach. **2**, 15–25 (2016)
21. Krol, O., Sokolov, V.: Modeling of spindle node dynamics using the spectral analysis method. In: Ivanov, V., Trojanowska, J., Pavlenko, I., Zajac, J., Peraković, D. (eds.) DSMIE 2020. LNME, pp. 35–44. Springer, Cham (2020). https://doi.org/10.1007/978-3-030-50794-7_4
22. Pavlenko, I., Trojanowska, J., Gusak, O., Ivanov, V., Pitel, J., Pavlenko, V.: Estimation of the reliability of automatic axial-balancing devices for multistage centrifugal pumps. Periodica Polytech. Mech. Eng. **63**(1), 52–56 (2019). https://doi.org/10.3311/PPme.12801
23. Ivanov, V., Dehtiarov, I., Pavlenko, I., Liaposhchenko, O., Zaloga, V.: Parametric optimization of fixtures for multiaxis machining of parts. In: Hamrol, A., Kujawińska, A., Barraza, M.F.S. (eds.) MANUFACTURING 2019. LNME, pp. 335–347. Springer, Cham (2019). https://doi.org/10.1007/978-3-030-18789-7_28

24. Kotliar, A., Basova, Y., Ivanova, M., Gasanov, M., Sazhniev, I.: Technological assurance of machining accuracy of crankshaft. In: Diering, M., Wieczorowski, M., Brown, C.A. (eds.) MANUFACTURING 2019. LNME, pp. 37–51. Springer, Cham (2019). https://doi.org/10.1007/978-3-030-18682-1_4
25. Krol, O., Sokolov, V.: Research of toothed belt transmission with arched teeth. Diagnostyka **21**(4), 15–22 (2020). https://doi.org/10.29354/diag/127193
26. Tsankov, P.: Modeling of vertical spindle head for machining center. J. Phys. Conf. Ser. **1553**, 012012 (2020). https://doi.org/10.1088/1742-6596/1553/1/012012
27. Kundrák, J., Molnár, V., Makkai, T., Dági, T.: Analysis of material removal efficiency in face milling of aluminum alloy. In: Gapiński, B., Szostak, M., Ivanov, V. (eds.) MANUFACTURING 2019. LNME, pp. 393–404. Springer, Cham (2019). https://doi.org/10.1007/978-3-030-16943-5_34
28. Ivanov, V., Liaposhchenko, O., Denysenko, Y., Pavlenko, I.: Ensuring economic efficiency of flexible fixtures in multiproduct manufacturing. Eng. Manag. Prod. Serv. **13**(1), 53–62 (2021). https://doi.org/10.2478/emj-2021-0004
29. Sokolov, V., Porkuian, O., Krol, O., Baturin, Y.: Design calculation of electrohydraulic servo drive for technological equipment. In: Ivanov, V., Trojanowska, J., Pavlenko, I., Zajac, J., Peraković, D. (eds.) DSMIE 2020. LNME, pp. 75–84. Springer, Cham (2020). https://doi.org/10.1007/978-3-030-50794-7_8
30. Krol, O., Porkuian, O., Sokolov, V., Tsankov, P.: Vibration stability of spindle nodes in the zone of tool equipment optimal parameters. Comptes rendus de l academie bulgare des sciences **72**(11), 1546–1556 (2019). https://doi.org/10.7546/CRABS.2019.11.12

Modification of the Spindle Head for a Milling Machine with Increased Load Capacity Drive

Oleg Krol(✉) and Volodymyr Sokolov

Volodymyr Dahl East Ukrainian National University, 59-a Tsentralnyi Prospect, Severodonetsk 93400, Ukraine

krolos@snu.edu.ua

Abstract. The design of a vertical spindle head of a multioperational machine tool with a modified gear drive is considered. 3D modeling of the spindle head in the integrated computer-aided design system KOMPAS-3D was carried out. An extended range of graphical primitives was used in the construction of complex geometry of the machine housing parts. Within the environment of the CAD system used, the express-construction procedure of high-precision models of gear rims with geometrically teeth correct surfaces was applied. A computational procedure for evaluating the geometric and energy parameters of a gear drive in module KOMPAS-GEAR is proposed. At the stage of choosing the tooth shape, it is possible to multivariate select the tooth axial shape. The calculation of stresses in the contact zone is carried out with the assumption of the carrying capacity ratio of the bevel wheel and the equivalent cylindrical wheel. A new idea to increase the load capacity of the bevel gear drive of the spindle head on the basis of modified generatrices of the gear and wheel initial surfaces is put forward. It is shown that the search for constructive improvement lies in the plane of increasing the teeth length of the contacting wheels. For the modified design of the gear transmission, analytical dependencies are determined for various geometric parameters at a constant angle of teeth inclination of the curvilinear generatrices of the cones. Numerical calculations have shown that the modified gear train is characterized by an increased length of teeth and a corresponding increased load capacity.

Keywords: Spindle unit drive · 3D modeling · Bevel gear

1 Introduction

Modular technological equipment is widely used as part of multioperational machine tools (MMT) of various layouts [1–3]. At the same time, various types of unified shaping units are being developed – horizontal, angular and vertical spindle heads [4–6]. On their basis, specialized multioperational machines with advanced technological capabilities are built, realizing the principle of maximum concentration of technological operations at one workplace.

The machine's layout differs in the spindle location, the main movement drives in space and the mutual arrangement of the main units. In the structure of MMT, the main

B. Gapiński et al. (Eds.): MANUFACTURING 2022, LNME, pp. 103–112, 2022.
https://doi.org/10.1007/978-3-031-00805-4_9

movement drive is particular importance and must implement high-productive execution of variety processes while ensuring a high quality of operational reliability.

In the drives of the machines main movement, a variant with an AC motor provided with a frequency transformer and a speed box based on 2 speed drives (with different types of gears) is often used to extend the range of velocity and momentum. This type of gear drive allows great torque for roughening and keepings high velocity finishing capability. High efficiency and reliability of machine tool drives depends on the resource and steady functioning of gears. Hence, the issues of increasing the durability, load carrying of the gear wheels by improving design solutions are one of the dominant areas of research and development.

In works [7], the need is observed at the beginning stages of investigation to increase durability to implement the finding for toolkit to characterize the contour of the side surface of the straight and bevel gear teeth. Such a definition will be the foundation for receiving under optimized contact requirements over a wide range of gear operation.

In the fundamental work [8], a new design of gears with increased loading capacity for contact stresses is proposed. The design component is based on material-physical relationships (in accordance with the accepted normative calculation ANSI/AGMA 2003 – B97) and allows you to evaluate the effect of the material on the subsurface layer. As for the geometric relationships, it should be noted that the new design of gears is characterized by a slight decrease in the normal modulus and the total overlap ratio with an increased average pitch diameter. Analysis of the contact stress criterion shows that an increase in the length of the working contact of the teeth increases the loading carrying of the gearing.

A constructive version of increasing the carrying capacity by the example of differential mechanisms is proposed in [9].

The substantive aspect related to the reduction of the length of contact of the teeth in the longitudinal direction in the process of engagement (in particular, at high cutting speeds) is considered in [10]. Such a change in the length of the loaded area of the teeth can lead to an unacceptable increase in contact stress. In this case, a generatrix of the single or dual spiral type is used as the shape of the tooth side surface for longer the contact distance. The manufacture of gears with such a complex geometrical shape of the tooth profile presents certain difficulties.

The procedure of three-dimensional representation of the designed mechanism is considered as an integrating link in the scale of the life cycle of creating a new gear mechanism. In accordance with [11], one of the major objectives is to generate a workable 3D model evaluation of the contact geometry, visible identity and assessment of the relationship of the variety of tooth geometric properties. Important parameters include the parameters of the groove shape, which are significantly defined by the shape of the lengthwise generatrix of the gear and wheel axoids.

The designer's toolkit should include a method for creating a solid model of a gear wheel in the appropriate integrated CAD systems [12, 13]. It is also important to provide the gear manufacturing with the appropriate CAM software. The Gleason 5-axis Gear Studio (G5S) software [14] from Gleason Corporation is promising in this regard. These customized programs are designed for profiling the side surface of the tooth, the contact drawing evaluation and the level of carrying capacity, as well as prepare appropriate control programs for five-axis CNC machines and processing centers.

On the basis analysis of the considered task of increasing the effectiveness of functioning and increasing the load capacity of different types of transmission, we formulate the problem:

Problem statement: Carry out study and search for the design of the MMT spindle head with an increased load capacity of the gear transmission for the machine tool drive.

To reach this objective, it is necessary to required the following problems:

1. To develop a three-dimensional model of the vertical spindle head of multioperational machine tools in the CAD KOMPAS-3D software using a specialized applied program "Shafts and mechanical transmissions-3D".
2. Research and develop a modernized design of a gear transmission with a modified form of the gear and wheel generating's line.

2 Three Dimensional Modeling of the Spindle Assembly

2.1 Construction of 3D Model of Vertical Spindle Head

As a design object, we consider the Vertical Spindle Head (VSH) of a specialized multifunctional machine tool of the 2nd standard size of milling type on the basis of the SVM1F4 design. The machine can operate with planes with the arbitrary arrangement in space, profile surfaces, bores, slots by different technological techniques: milling, drilling, countersinking and reaming.

A special spindle head realizes the main shaping movement using a variable drive using a DC engine and a thyristor stress transformer. To analyze the functionality of the construction and choose the optimal design of the spindle head for a customized multitasking CNC machine tool SVM1F4, its 3D model was made in the computer-aided design KOMPAS-3D software [15–17] (Fig. 1).

The system includes a whole set of software tools for developing such a rather complex VSH-design, which includes, first of all, the C3D Modeler geometric core and the C3D Solver parametric core. The C3D Modeler geometry kernel performs the construction of the geometric model and the calculation of the geometric characteristics of the modelled object. The new version of C3D Modeler we are using now has functionality for removing bores and fillets (Fig. 2). This leads to a simplification of the 3D model, which is prepared for further calculation in the CAE system.

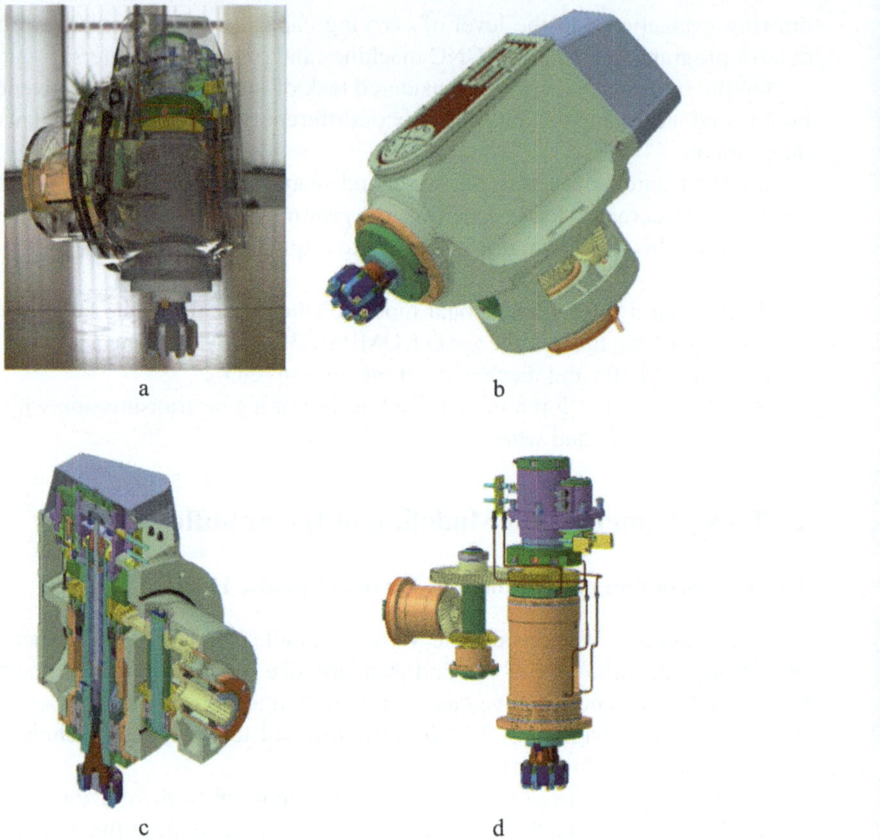

Fig. 1. Solid model of the Vertical Spindle Head: a – VSH transparent representation; b – WSH-assembly; c – WSH-cross-section; d – 3D-kinematics

2.2 3D Modeling of Housing Part

Toolkit and a set of KOMPAS graphic primitives allow you to effectively create three-dimensional models of complex housing products (Fig. 2), as the basis of the carrying system of CNC milling machines [18, 19]. The KOMPAS-3D system has basic operations: extrusion, rotation, kinematic operation and section operations for constructing volumetric elements and surfaces. This system allows you to implement the strategy of creating 3D models based upon the forming internal method for future solid models using the “Excision” operation to form cavities and channels of complex shapes.

2.3 3D Modeling of Bevel Gears

Solid models of bevel gears are constructed with the “Shafts and mechanical transmissions-3D” application. On the basis of this application, it becomes possible to operate not with geometric primitives of the plane or elements of space, but with objects and details in general, as well as their elements. An important advantage here

a b

Fig. 2. The housing of the spindle head: a – general view; b –section

is the ability to quickly build high-precision models of gear rims with geometrically correct tooth surfaces. The speed of constructing complex surfaces has increased due to the introduction of a unique technique for constructing gear rims by the method of imitation of gear milling of bevel gears [13, 20, 21].

This plan also uses the new CAD KOMPAS-3D functionality, which works with a new type of surface "Conical section surface", which is formed by moving a conical section curve along two guides with the ability to change the parameters of this section. As a result, a very smooth surface is formed along its entire length. In Fig. 3 shows 3D models of a bevel wheel and pinion.

a b

Fig. 3. Solid models of bevel gear elements: a – gear wheel; b – pinion

3 Design of Bevel Gears with Circular Teeth in the KOMPAS-GEAR Module

Consider the design procedures for bevel gears of a spindle head with a circular tooth in the KOMPAS_GEARS mechanical transmission calculation module, it is possible to calculate bevel gears with circular teeth of external engagement with center-to-center angles from 10° to 150°, a tooth tilt angle from 0° to 45° [22–24].

Since the production of the spindle heads under consideration is characterized by a small-scale production type, the main calculation option is a one-dimensional cutting method with options for one- and multi-sided machining [25–27].

The basic version of the bevel gear is characterized by the following initial data: torque on the wheel $T_2 = 350$ N · m; gear ratio $n_1 = 218$ min^{-1}.

A preliminary calculation of the basic version showed that the main parameters take on the following values: average normal modulus $m_{nm} = 3.52$ mm; average conical distance $R_m = 76.52$ mm; inclination angle of the pitch midline of the tooth $\beta_{nm} = 35^{\circ}$. In this case, the second type of axial shape of teeth is chosen – lowering teeth in which the tops of the pitch cones and roots do not coincide. The results of calculating the geometric parameters for the KOMPAS GEARS module are illustrated in Fig. 4, a.

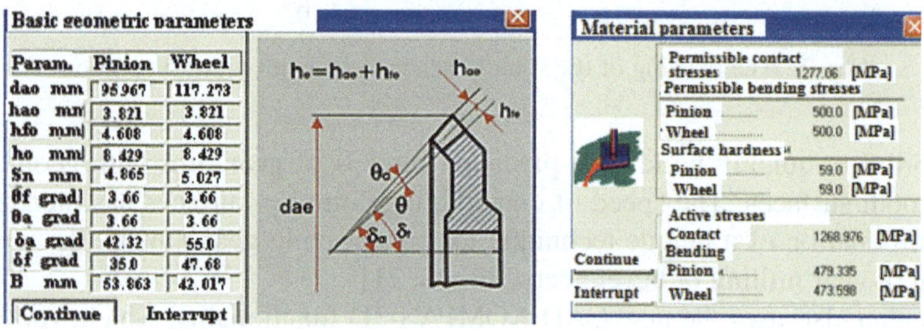

a b

Fig. 4. Bevel gear design with a circular tooth: a – the wheel geometry; b – stress in contact

The module allows you to efficiently calculate the energy parameters [28, 29] – stresses in the contact zone (Fig. 4, b) The main assumption is that the carrying capacity of the conical gear teeth is identical an equivalent cylindrical gear with the same tooth length b and a profile corresponding to the middle additional cone (middle section of the tooth).

As a result of calculating the energy parameters in the KOMPAS GEARS module, the quantitative values of the cutting forces components and the permissible bending stresses $\sigma_{FP\,\max} = 500$ MPa; as well as the permissible contact stresses $\sigma_{HP\,\max} = 1277.0$ MPa were obtained. Calculations of the effective contact stress $\sigma_H = 1268.98$ MPa have shown that its value approaches the maximum permissible, which makes it expedient to search for a method that decreases the level of area stresses in the contact zone.

4 Design Method for Increasing the Load Capacity of the Gear

Along with the technological methods of increasing exploitation properties (hardening of the outer surfaces by heat treatment, wear-resistant coatings), the design methods of improving the transmissions of the machine drive have not been exhausted.

There is an integral indicator of the effective operation of the gear – the load capacity of the tooth surface. In bevel gears [30], the load capacity is limited by the area of the contact surfaces, which depends on the length of the pinion teeth. In turn, this size depends on the ratio of the rectilinear generatrix directions of the gear and wheel axoids on the one hand and the direction of the teeth on the other side. As a consequence, the teeth length is limited by the width of the gear engagement.

This paper proposes a constructive improvement of the bevel gear. In a modified version of the gear, this is achieved by increasing the teeth length, due to which there will be a higher load capacity of the transmission. This increase is achieved by the fact that in a modified bevel gear, the generatrix of the gearing initial surfaces is curved and represents a circular arc radius R (Fig. 5).

The radius of the curvilinear generatrix is determined by the relationship (1):

$$R = \frac{\Delta + 0.5 \cdot b \cdot \cos \delta_{m1}}{\sin \delta_{m1}}, \tag{1}$$

where $\Delta = 4 \cdot m_{tm}$; m_{tm} – middle end gearing module; δ_{m1} – the average pitch.

The teeth of the gear and the wheel are located at an angle β to the curved generators of the initial surfaces of the gear and wheel. In this bevel gear transmission with curved generatrices of the initial surfaces radius of gear 1 and wheel 2, oblique teeth are cut on the gear rims in width b (Fig. 5, a).

The teeth inclination angle β of the cones curved generatrices for gear 1 and wheel 2 is constant in the entire direction from the outer diameters d_{e1} and d_{e2} to the inner diameters d_{i1} and d_{i2}. Its value lies within: $\beta = 25^\circ \ldots 35^\circ$.

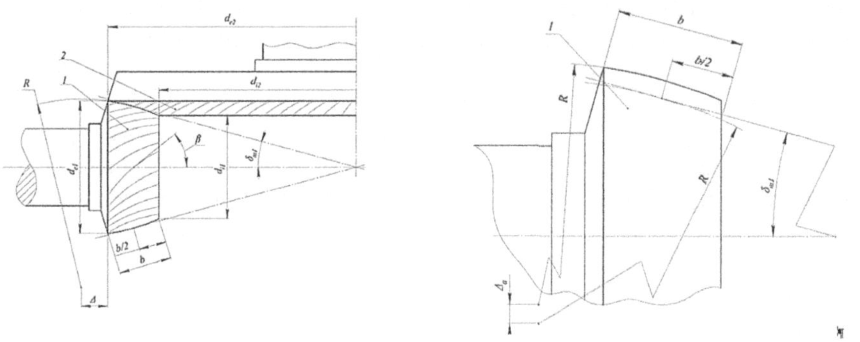

Fig. 5. Modified gear with curved generating line: a – new transmission; b – engagement zone

In Fig. 5, b displays a section of a workpiece for a bevel gear 1. The surface of the gear workpiece (in Fig. 5, b it is shown with a dash-dotted line) coincides with the surface of the teeth tops of gear 1 and is outlined, like the initial surface, by an arc of circle R. The centers of these arcs are displaced from each other by the value Δ_a (2):

$$\Delta_a = m_{tm} / \cos \delta_{m1}. \tag{2}$$

The initial surface of the wheel 2 is formed, as in the gear 1, by a circle arc of radius R. As a result of the teeth inclination, the bevel gear transmission provides rotation from gear 1 to wheel 2 with improved technical indicators – increased load capacity and smoother engagement. Calculations show that for the indicated values of the angle β : $\beta = 25^\circ \ldots 35^\circ$ the length of the teeth and their contact lines is (10… 22) % greater than in the known transmission. As a result, the bevel gear has an advantage over known gear designs in terms of load capacity.

5 Conclusion

1. A complex of the vertical spindle head 3D models for a multioperational milling machine with a gear drive in the computer-aided design KOMPAS-3D software has been developed. The strategy of creating solid models based on the forming by internal method with using the "Excision" has been effectively used.
2. In the specialized application "Shafts and mechanical transmissions 3D", three-dimensional models of bevel gears are created. The effectiveness of using this application is due to the introduction of a unique procedure for constructing gear rims by the method of imitation of gear-tooth milling for bevel gears.
3. A complex calculation of bevel gear with circular teeth has been implemented in a specialized module KOMPAS-GEAR. The main geometrical and power parameters (forces and stresses) have been determined when choosing as initial data.
4. The constructive improvement of the bevel gear transmission by choosing the arc of a circle as the generatrix of the gear and the wheel initial surfaces is implemented. Analytical dependencies for determining the radii R and teeth inclination angles β of the curvilinear generators are given. The results obtained in the work make it possible to increase the length of the teeth and their contact lines by (10… 22) % more than in the known gear, which gives an increase in the load capacity.
5. Research in the field of creating new and modifying successfully operated spindle heads of multi-operation milling machines and their drives have shown a significant reserve in improving the designs of mechanical transmissions and increasing the service life. As a further direction of improvement, the development of parametric tools is considered as a means of modeling an extended range of spindle heads and the creation of specialized sections of databases that are embedded in the contour of modern CAD systems.

References

1. Brecher, C., Fey, M., Daniels, M.: Modeling of position-, tool- and workpiece-dependent milling machine dynamics. High Sped Mach. **2**, 15–25 (2016). https://doi.org/10.1515/hsm-2016-0003
2. Brecher, C., Bäumler, S., Daniels, M.: Prediction of dynamics of modified machine tool by experimental substructuring. In: Allen, M., Mayes, R., Rixen, D. (eds.) Dynamics of Coupled Structures, vol. 1. Conference Proceedings of the Society for Experimental Mechanics Series. Springer, Cham (2014).https://doi.org/10.1007/978-3-319-04501-6_28
3. Gao, X., Li, B., Hong, J., Guo, J.: Stiffness modeling of machine tools based on machining space analysis. Int. J. Adv. Manuf. Technol. **86**(5–8), 2093–2106 (2016). https://doi.org/10.1007/s00170-015-8336-z
4. Kong, J., Cheng, X.: Modal analysis of CNC lathe's spindle based on finite element. J. Adv. Eng. Res. (AER) **148**, 318 (2017)
5. Xi, S., Cao, H., Chen, X., Niu, L.: Dynamic modeling of machine tool spindle bearing system and model based diagnosis of bearing fault caused by collision. Procedia CIRP **77**, 614–617 (2018). https://doi.org/10.1016/j.procir.2018.08.197
6. Abele, E., Altintas, Y., Brecher, C.: Machine tool spindle units CIRP. Ann. Manuf. Technol. **59**, 781–802 (2010)

7. Kushnir, E., Portman, V.T., Aguilar, A., Clark, W.: Layout evaluation at earlier stages of machine tool design: form-shaping function-based approach. Int. J. Adv. Manuf. Technol. **90**(9-12), 3333–3346 (2016). https://doi.org/10.1007/s00170-016-9667-0
8. Wirth, C., Höhn, B-R., Braykoff, C.: New methods for the calculation of the load capacity of bevel and hypoid gears. In: Gear technology, June/July, pp. 44–54 (2013)
9. Malashchenko, V., Strilets, O., Strilets, V., Kłysz, S.: Investigation of the energy effectiveness of multistage differential gears when the speed is changed by the carrier. Diagnostyka **20**(4), 57–64 (2019). https://doi.org/10.29354/diag/112397
10. Amendola, J.B., Amendola III, J.B., Yatzook, D.: Longitudinal tooth contact pattern shift. Geartechnology **2012**(May) 63–67 (2012)
11. Bjionowski, B.: A practical approach for modeling a bevel gear. Geartechnology 2015 March/April, pp. 68–75 (2015)
12. Nikonov, V.V.: KOMPAS 3D. Creation of Models. Peter, St. Petersburg (2020)
13. Chagina, A.V., Bolshakov, V.P.: 3D Modeling in KOMPAS-3D v17 and Higher Textbook for Universities. Peter, St. Petersburg (2021)
14. Gaiser, U.: 5-axis gear manufacturing. Geartechnology **2017**(March), 32–34 (2017)
15. Tsankov, P.: Modeling of vertical spindle head for machining center. J. Phys. Conf. Ser. **1553**, 012012 (2020) https://doi.org/10.1088/1742-6596/1553/1/012012
16. Ivanov, V., Dehtiarov, I., Pavlenko, I., Liaposhchenko, O., Zaloga, V. Parametric optimization of fixtures for multiaxis machining of parts. In: Hamrol, A., Kujawińska, A., Barraza, M. (eds.) Advances in Manufacturing II. MANUFACTURING 2019, vol. 2, pp. 335–347. LNME. Springer, Cham (2019).https://doi.org/10.1007/978-3-030-18789-7_28
17. Demianenko, M., et al.: Impact of dynamic characteristics of gears on the reliability of prilling equipment. In: Knapcikova, L., Peraković, D., Perisa, M., Balog, M. (eds.) Sustainable Management of Manufacturing Systems in Industry 4.0. EAI/SICC. Springer, Cham (2022). https://doi.org/10.1007/978-3-030-90462-3
18. Pavlenko, I., Trojanowska, J., Gusak, O., Ivanov, V., Pitel, J., Pavlenko, V.: Estimation of the reliability of automatic axial-balancing devices for multistage centrifugal pumps. Periodica Polytech. Mech. Eng. **63**(1), 52–56 (2019). https://doi.org/10.3311/PPme.12801
19. Platonov, L.: Machine-building design in KOMPAS-3D at a new round of development. CAD Graph. **10**, 6–12 (2013)
20. Krol, O., Sokolov, V.: Research of modified gear drive for multioperational machine with increased load capacity. Diagnostyka **21**(3), 87–93 (2020). https://doi.org/10.29354/diag/126026
21. Demianenko, M., Volf, M., Pavlenko, I., Liaposhchenko, O.: Experimental studies on oscillation modes of vibration separation devices. J. Eng. Sci. **8**(1), D1–D9 (2021). https://doi.org/10.21272/jes.2021.8(1).d1
22. Rogovyi, A., Korohodskyi, V., Khovanskyi, S., Hrechka, I., Medvediev, Y.: Optimal design of vortex chamber pump. J. Phys. Conf. Ser. **1741**(1), 012018. IOP Publishing (2021). https://doi.org/10.1088/1742-6596/1741/1/012018
23. Gasanov, M., Kotliar, A., Basova, Y., Ivanova, M., Panamariova, O.: Increasing of lathe equipment efficiency by application of gang-tool holder. In: Gapiński, B., Szostak, M., Ivanov, V. (eds.) Advances in Manufacturing II. MANUFACTURING 2019, vol. 4, pp. 133–144. LNME. Springer, Cham (2019). https://doi.org/10.1007/978-3-030-16943-5_12
24. Pavlenko, I., Trojanowska, J., Ivanov, V., Liaposhchenko, O.: Parameter identification of hydro-mechanical processes using artificial intelligence systems. Int. J. Mechatron. Appl. Mech. **2019** (5), 19–26 (2019)
25. Verbovyi A., et al.: Parameter identification of nonlinear bearing stiffness for turbopump units of liquid rocket engines considering initial gaps and axial preloading. J. Eng. Sci. **8**(2), D8-D11 (2021). https://doi.org/10.21272/jes.2021.8(2)

26. Pacana, J., Kozik, B., Budzik, G.: Strength analysis gears in dual path gearing by means of FEM. Diagnostyka **16**(1), 41–46 (2015)
27. Shevchenko, S., Mukhovaty, A., Krol, O.: Gear transmission with conic axoid on parallel axes. In: Radionov, A., Kravchenko, O., Guzeev, V., Rozhdestvenskiy, Y. (eds.) Proceedings of the 5th International Conference on Industrial Engineering (ICIE 2019). ICIE 2019. LNME. Springer, Cham (2020).https://doi.org/10.1007/978-3-030-22041-9_1
28. Fomin, O., Lovska, A.: Establishing patterns in determining the dynamics and strength of a covered freight car, which exhausted its resource. Eastern-Eur. J. Enterp. Technol. **6**(7(108)), 21–29 (2020) http://journals.uran.ua/eejet/article/view/217162
29. Krol, O., Porkuian, O., Sokolov, V., Tsankov, P.: Vibration stability of spindle nodes in the zone of tool equipment optimal parameters. Comptes rendus de l academie bulgare des sciences **72**(11), 1546–1556 (2019). https://doi.org/10.7546/CRABS.2019.11.12
30. Kačalova, M., Pavlenko, S.: Strength and dynamic analysis of a structural node limiting the multi-output gear mechanism. Acta Polytechnika **57**(5), 316–320 (2017)

Theoretical Model of the Densification During Hot Pressing and its Verification

Edwin Gevorkyan[1], Miroslaw Rucki[2(✉)], Volodymyr Nerubatskyi[1], Zbigniew Siemiatkowski[2], Dmitrij Morozow[2], and Hanna Komarova[1]

[1] Ukrainian State University of Railway Transport, Kharkiv, Ukraine
[2] Kazimierz Pułaski University of Technology and Humanities in Radom, Radom, Poland
m.rucki@uthrad.pl

Abstract. In the paper, theoretical model of the densification process was proposed for the sintering under pressure and directly applied electric current. Despite the wide application of sintering processes in industry, it is still discussed what mechanisms are responsible for densification and how to predict the properties of the sintered material. In this paper, the model was proposed based on summation of the partial densities obtained for the given temperature and pressure from plastic deformation, diffusion, power law creep, Nabarro-Herring creep during the process duration time. The process itself was divided to two stages, before the relative density 0.9 is reached, and after it. Initial experimental results proved satisfactory conformity with the model below 10%.

Keywords: Densification · Sintering · Hot-pressing · Creep · Model

1 Introduction

In a powder metallurgy, full density processing techniques based on hot pressing are widely applied. A group of novel consolidation techniques that combine the advantages of vacuum sintering and hot pressing include pressure-assisted densification and unique benefits from the direct application of electric current [1]. In general, electroconsolidation is advantageous because of the lower sintering temperature, higher heating rate, and shorter sintering time than that of other sintering techniques. Electrical field has certain effects on mass transport, reactivity, diffusion, grain boundary mobility and other processes that take place during compaction of the powder [2]. However, there are some difficulties with modeling of the process, since early presumptions on the role of plasma and discharge phenomena in the field assisted sintering, that even resulted in designation "spark plasma sintering," was probably wrong [3].

Other problem with theoretical description of the densification is related to the dimensions of particles. The sintering mechanism in the case of micrometer-sized particles is well studied and many papers have been published, while the sintering of submicron powders has its own specifics, where micromechanistic models for densification are not applicable [4]. Despite variety of the theoretical explanations, there is no comprehensive model that would describe the entire process.

B. Gapiński et al. (Eds.): MANUFACTURING 2022, LNME, pp. 113–123, 2022.
https://doi.org/10.1007/978-3-031-00805-4_10

In the current research, theoretical model for the hot-pressing under directly applied electrical current is proposed. Material of the simulations and experiments was chosen aluminum oxide (Al_2O_3), since it is "one of the most important ceramic materials and its dense nanocrystalline form (with <100 nm grain size) is of great scientific and technological interest" [5]. Densely sintered alumina exhibit outstanding wear resistance and strength, valuable in many mechanical applications [6], as well as nanoporous alumina-based ceramics [7]. Thus, the investigations were aimed at finding reliable theoretical model useful also for submicron alumina powders.

2 Theoretical Model

2.1 General Assumptions

Model of the hot-pressing consolidation under directly applied electrical current of 50 Hz was based on the classical theory described in the works by Ashby and co-authors [8, 9]. Essentially, the model assumes possibility of simultaneous processes of plastic deformation, power law creep, Nabarro-Herring creep, as well as mass transfer as a result of grain-boundary diffusion and surface diffusion. The powder was described as a set of spherical particles of similar diameters [10], randomly compacted. Theoretically, if the sample was fully compacted with no pores, the bulk density ρ would be reached. Since the pores are not fully eliminated, the relative density D_R of the sample is smaller than 1. It is usually convenient to distinguish between three stages of sintering, namely initial, intermediate and final ones [11]. However, in the present model, it was useful to divide the densification process to two stages only: first stage covered relative density D_R from the loose powder initial density D_{R0} up to $D_R = 0.9$, while the second stage dealt with relative densities $0.9 < D_R < 1.0$. The densification was understood as a decrease of the overall volume due to shrinkage of the distances between particles and, thus, was modeled as increase of the single particle radiuses around fixed centers. Number of contacts between a single particle and other particles is calculated, and then mean contact surface, radius of the neck surface is determined to find the effective pressured in each contact point. In the second stage, the surface of a particle takes form of a polyhedron and the effective pressure decreases because of inner pressure of the pores concentrated in the corners of the polyhedrons.

The relative density was determined through summation of the partial densities obtained for the given temperature and pressure from plastic deformation, diffusion, power law creep, Nabarro-Herring creep during the process duration time.

As a result, it was possible to plot the diagrams of hot-pressing, i.e. dependence between the relative density D_R and process parameters (pressure, temperature, time). The subjects of investigation were powders of alumina of average particle diameters 0.6 μm. Next, series of experiments were performed to obtain the dense samples out of these powders and to compare the results with theoretical model. The applied pressure was between 20 and 45 MPa and process duration from 5 up to 30 min. It was found that the model did not provide reliable results in this range of pressures, temperatures and Al_2O_3 powder dimensions. In general, experiments provided much higher densification rate than that calculated, so it was necessary to modify the model.

2.2 Initial Stage of Densification

It is widely admitted that the origin of the ultra-rapid densification in the electrical field is still controversial [12]. It is known that various mass transfer mechanisms depend significantly on the temperature, pressure, and certain dimensions of the diffusion ways. Especially, it is important to consider a linear dimension for nanosize powders hot-pressing, where diffusion processes play the key role in the mass transfer [13]. Moreover, under the outer electrical field, e.g. during spark plasma sintering, diffusion coefficients considerably increase [14, 15]. It can be assumed that in the first stage of densification, when relative density is $D_{RO} < D_R < 0.9$, initially spherical particles join together increasing number of contacts with intensely growing neck radiuses. Around a fixed particle center of initial radius R, other particles are joining causing the growth of a solidified particle and increase of its radius to the new dimension R'. The formulas derived from the Ashby's model [8, 9] in application to this particular case, can be presented as follows:

$$R' = \left(\frac{D_R}{D_{R0}}\right)^{1/3} R, \tag{1}$$

where D_{R0} and D_R are initial and actual density, respectively. For the random compaction of the particles in the powder, it can be assumed $D_{R0} = 0.64$.

When the spherical particle grows, coalescence of two spheres takes place by surface diffusion. As the distance between their two centers decrease, the interconnection volume between two spheres can be calculated and redistributed to the empty spaces. At the same time, neck radiuses increase and new contacts between particles are initiated increasing their number Z. Dependence between the number of contacts with neighboring spheres Z and the relative density D_R can be written as follows:

$$Z = 12D_R, \tag{2}$$

In fact, number of contacts Z increases from 7.7 at the beginning of hot isostatic pressing, when relative density of the loose powder is $D_{RO} = 0.64$, up to $Z = 12$ when the full density $D_R = 1$ is reached. Average contact surface s_c can be calculated from the following equation:

$$s_c = \frac{\pi}{3}\frac{(D_R - D_{R0})}{(1 - D_{R0})}R^2, \tag{3}$$

The radius of the neck r_n is dependent on the relative density D_R, and this dependency can be expressed as follows:

$$r_n = \sqrt{\frac{s_c}{\pi}} = \frac{1}{\sqrt{3}}\left(\frac{D_R - D_{R0}}{1 - D_{R0}}\right)^{1/2} R. \tag{4}$$

On the other hand, the curvature radius of the neck γ_n can be calculated as follows:

$$\gamma_n = \frac{r_n^2}{2(R - r_n)} \cong R(D_R - D_{R0}). \tag{5}$$

Under the outer pressure P, contact areas between particles are subject to the force f, which in average can be calculated as follows:

$$f = \frac{4\pi R^2}{Z} P. \tag{6}$$

Hence, effective pressure P' in each point can be determined from the following equation:

$$P' = \frac{f}{s_c} = \frac{(1 - D_{R0})}{D_R{}^2 (D_R - D_{R0})} P. \tag{7}$$

It can be noted that P' would be close to P when D_R is close to 1, i.e. when the full theoretical density is reached.

However, during the second stage of the sintering process, when $0.9 < D_R < 1$, the empty spaces are trapped between the particles and form closed pores. It can be assumed that the pores are distributed in the corners of the polyhedron-form particles. Radius of the pores r_p can be calculated as follows:

$$r_p = \left(\frac{1 - D_R}{6}\right)^{1/3} R. \tag{8}$$

The closure of the pores poses a problem with their inner pressure P_i, which prevents sintered material from reaching its full theoretical density. The inner pressure can be calculated as follows:

$$P_i = P_0 \frac{(1 - D_{RC}) D_R}{(1 - D_R) D_{RC}}, \tag{9}$$

where P_0 is the air pressure in the pores at the closure moment, and D_{RC} is a critical value of relative density that corresponds with the pores closure.

2.3 Relative Density Summation

In the model, three densification mechanisms were considered, namely plastic deformation, diffusion and creep. Summation of all these phenomena provides overall understanding of the relative density rate.

Plastic deformation works after certain pressure $P' > 3\sigma_f$ is reached, where σ_f is the flow stress of the material. Simultaneously, the contact area between particles is growing. In the first stage of sintering, densification promoted by plastic deformation can be described by the following equation:

$$D_{Rpl} = \left(\frac{(1 - D_{R0}) P}{1.3\sigma_f} + D_{R0}{}^3\right), \tag{10}$$

where D_{Rpl} is the relative density generated by plastic deformation. However, in the second stage of sintering, D_{Rpl} is described by other equation:

$$D_{Rpl} = 1 - \exp\left(-\frac{3}{2}\frac{P}{\sigma_f}\right). \tag{11}$$

In the cases of *diffusion* and *creep* mechanisms, it was found convenient to describe them using densification rates $\dot{D}_{R1}$ and $\dot{D}_{R2}$, respectively. Again, diffusion rate during the first stage of sintering is calculated differently from the one during the second stage. In the first stage, mass transfer tales place from the contact zones between powder particles towards the neck formed due to the grain-boundary and volume diffusion. Thus, in the first stage, diffusion-promoted densification rate $\dot{D}_{R1}$ can be calculated as follows:

$$\dot{D}_{R1} = \frac{43(1 - D_{R0})^2}{(D_R - D_{R0})^2} \frac{(\delta D_b - \gamma_n D_v)}{kTR^3} \Omega P, \tag{12}$$

where D_v and D_b are the diffusion coefficients by volume and by grain-boundaries, respectively, δ is the effective thickness of a grain boundary, k is the Boltzmann constant, T is the absolute temperature, and Ω is the atomic volume. However, in the second stage of the sintering process, it should be calculated from different formula, considering radiuses of pores r_p instead of radiuses of necks γ_n:

$$\dot{D}_{R1} = 54 \frac{\Omega(\delta D_b - r_p D_v)}{kTR^3} P \sqrt[5]{1 - D_R}. \tag{13}$$

Generally, creep can be described by the following formula:

$$\dot{\varepsilon} = \dot{\varepsilon}_0 \left(\frac{\sigma}{\sigma_0} \right)^n, \tag{14}$$

where $\dot{\varepsilon}$ is deformation rate, σ is the actual stress, while $\dot{\varepsilon}_0$, σ_0, and n are the parameters of the particular material.

However, two types of creep should be distinguished in the model. It is a deformation-promoted creep with related densification rate $\dot{D}_{R2}$ in the areas of contact between the particles, on one hand, and Nabarro-Herring and Coble creep with distinctive densification rate $\dot{D}_{R3}$, on the other hand. Both types should be calculated differently in the first and the second stages of the sintering process.

The densification rate $\dot{D}_{R2}$ related to the first type of creep in the first stage of sintering, when relative density is below 0.9, can be calculated as follows:

$$\dot{D}_{R2} = \frac{3}{2} \left(\frac{\dot{\varepsilon}_0}{\sigma_0^n} \right) \frac{D_R(1 - D_R)}{(1 - [1 - D_R]^{1/n})^n} \left(\frac{3}{2n} P \right)^n. \tag{15}$$

However, when the relative density is above 0.9 and is approaching 1, densification rate should be calculated as follows:

$$\dot{D}_{R2} = 5.3 \left(D_R{}^2 D_{R0} \right)^{1/3} \frac{r_n}{R} \left(\frac{\dot{\varepsilon}_0}{\sigma_0^n} \right) \left(\frac{P'}{3} \right)^n. \tag{16}$$

On other hand, Nabarro-Herring and Coble creep mechanisms appear in the lowest stress range [16]. They cannot occur in monocrystalline powder particles without grain boundaries, and thus, are typical for the pressure-assisted sintering mechanism [17]. These mechanisms take place when the grain dimensions are substantially smaller than that of powder particles. During the first stage of sintering process, the following formula can be applied to calculate densification rate $\dot{D}_{R3}$ related to the Nabarro-Herring and Coble creep mechanisms:

$$\dot{D}_{R3} = 24.9\frac{\Omega}{kT\overline{G}^2}\left(D_R^2 D_{R0}\right)^{1/3}\frac{r_n}{R}\left(D_v + \frac{\pi\delta D_b}{\overline{G}}\right)P', \tag{17}$$

where $\overline{G}$ is an average grain size. However, when the relative density is above 0.9 and is approaching 1, densification rate $\dot{D}_{R3}$ should be calculated as follows:

$$\dot{D}_{R3} = 31.5\frac{\Omega}{kT\overline{G}^2}(1 - D_R)^{1/3}\frac{r_n}{R}\left(D_v + \frac{\pi\delta D_b}{\overline{G}}\right)P. \tag{18}$$

Finally, the relative density can be obtain from the summation of the partial densification promoted by plastic deformation and the three respective densification rates multiplied by the pressing time t, as follows:

$$D_R = D_{Rpl} + \sum_{i=1}^{3} \dot{D}_{Ri}t. \tag{19}$$

2.4 Algorithm for Alumina Powder Calculations

In order to plot the diagrams for densification during hot isostatic pressing of alumina, the main physical data related to the sintering process were taken from literature [18] and given in Table 1.

Table 1. Chosen physical and mechanical properties of Al_2O_3.

	Property, units	Value
1	Atomic volume, Ω (m^3)	$4.25 \cdot 10^{-29}$
2	Burgers vector, b (m)	$4.76 \cdot 10^{-10}$
3	Melting point, T_m (K)	2320
4	Flow stress, σ_f (MPa)	1500
5	Shear modulus at 300 K, μ_0 (MPa)	$0.24 \cdot 10^4$
6	Ratio $\frac{T_m d\mu}{\mu_0 dT}$	– 0.35
7	Volume diffusion, D_{0v} (m^2/s)	$2.8 \cdot 10^{-10}$
8	Grain-boundary diffusion, δD_{0b} (m^3/s)	$8.6 \cdot 10^{-10}$

Based on the data from Table 1 and the equations from (1) to (19), the alumina powder densification process under directly applied electrical current was simulated. The calculation algorithm is presented in Fig. 1.

Apart from the equations from (1) to (19), it was found necessary to introduce corrections. Some effect on the final results may have contaminations on the surface of the

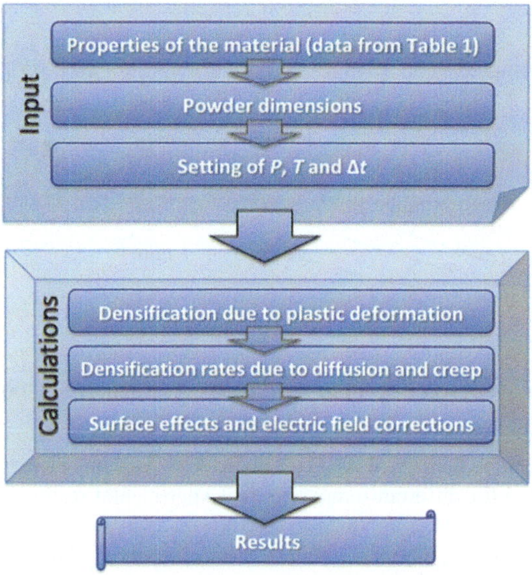

Fig. 1. Algorithm of calculation of alumina sintered sample relative density under directly applied electric current.

powder particles, as well as their difference from the spherical form and non-uniform dimensions. Some corrective coefficients were introduced, accordingly. Moreover, deviations of the electric field from the theoretical values also were taken into consideration in the calculations.

3 Experimental Verification

To verify the model, several experiments were performed using the equipment for field activated sintering [19]. Time and resources constrains did not allow for full statistical verification, but interesting initial results were obtained for different pressures. Both experimental results and theoretical data are shown in Figs. 2, 3 and 4, where colored plots correspond with calculations, and black points represent experimental data.

In Fig. 2, there is a plot of theoretical relative density for different sintering temperatures, calculated for the pressure 20 MPa. In accordance with expectations, the model shows that higher density can be obtained at certain temperature when the holding time is longer. It is noteworthy, however, that the time $t = 5$ min provided almost linear function $D_R = f(T)$. Since 20 MPa is rather rarely used pressure, only one experiment was performed for this value of P at temperature 1290 °C. The relative density $D_R = 0.84$ was obtained, which was much closer to the theoretical value for 1 min (0.83) than that for 5 min (0.90).

Nevertheless, difference of 7% between theoretical and experimental values for the temperature 1290 °C at pressure $P = 20$ MPa and pressing time 5 min can be considered satisfactory.

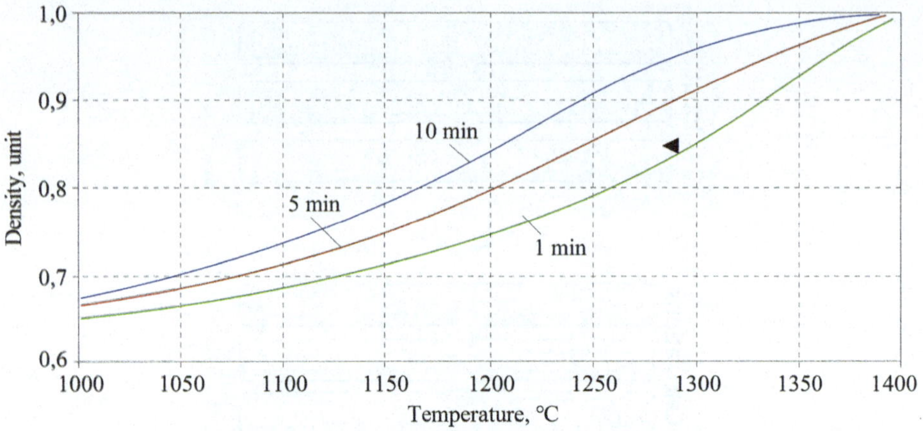

Fig. 2. Theoretical relative density of Al_2O_3 powder with particles diameter 0.6 μm at pressure $P = 20$ MPa calculated for different temperatures. Experimental result: ◀– $P = 20$ MPa, $T =$ 1290 °C, $t = 5$ min, $D_R = 0.84$.

Figure 3 presents a graph of theoretical relative density for different sintering temperatures, calculated for the pressure 45 MPa. It should be noted that the general trend is the same as for 20 MPa (Fig. 2), but the curve corresponding with holding time $t =$ 5 min exhibited less linearity. At the temperatures above 1300 °C, increase of density is smaller.

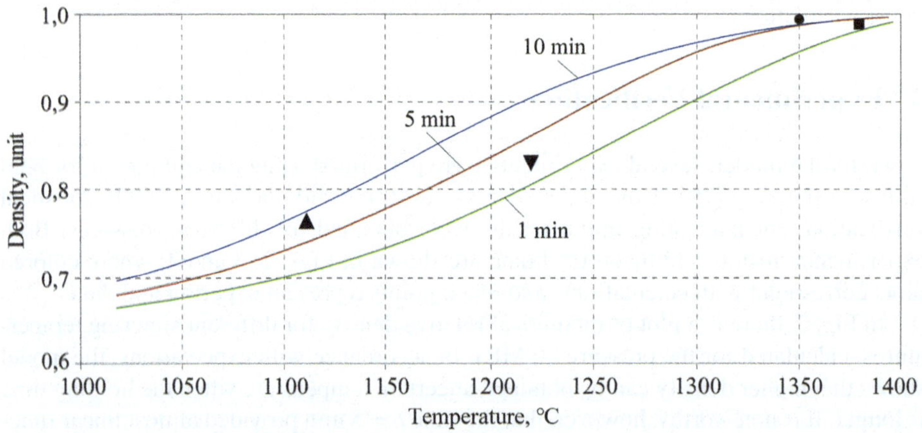

Fig. 3. Theoretical relations between relative density of hot-pressed Al_2O_3 powder (particles diameter 0.6 μm) and sintering temperature at the pressure $P = 45$ MPa, and experimental results for pressing time 5 min: ▲– at temperature 1110 °C, obtained relative density $D_R = 0.766$; ▼– at 1220 °C, $D_R = 0.83$; • – at 1350 °C, $D_R = 0.999$; ■– at 1380 °C, $D_R = 0.999$.

Important characteristic can be derived from the Fig. 3 is that at temperature 1330 °C, the relative density obtained after 5 min will not improve after 10 min. In fact, no substantial improvement can be achieved also through further increase of the temperature, since experimental $D_R = 0.999$ was obtained both at 1350 °C and at 1380 °C.

In the case of lower sintering temperatures, experimental results exhibited good conformity with the theoretical model. The difference was less than ±4% between experimental and calculated relative densities D_R.

Keeping in mind that the nanopowders could provide more dense material and nanostructural ceramics posses better characteristics than the microstructural ones [20], additional experiment was performed with alumina nanopowder. The particle dimensions were 0.06 μm, which theoretically was expected to provide much higher relative density at the same temperatures and holding times, as it can be seen in Fig. 4. The experimental point of $D_R = 0.999$ was obtained after 5 min of sintering at the temperature 1250 °C.

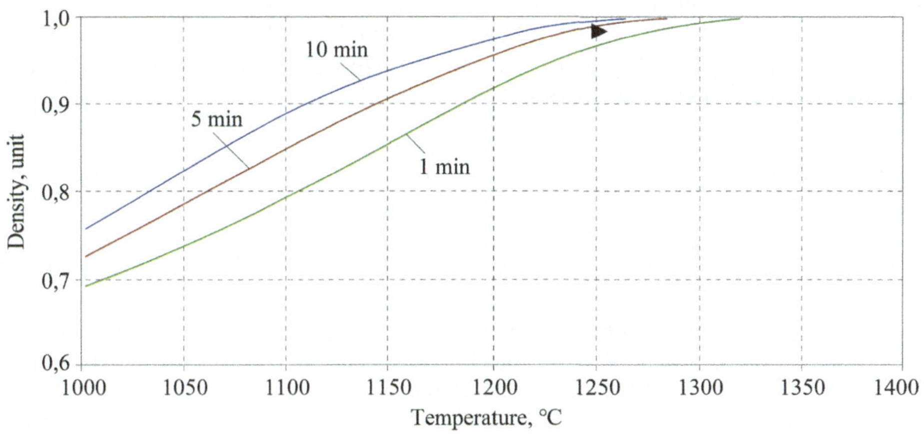

Fig. 4. Theoretical relative density of Al_2O_3 nanopowder with particles diameter 0.06 μm at pressure $P = 45$ MPa calculated for different temperatures. Experimental: ▶– $P = 45$ MPa, $T =$ 1250 °C, $t = 5$ min, $D_R = 0.999$.

From the practical perspective, these results are enough to confirm the fidelity of the proposed model. It makes possible to estimate the temperature and holding time for the alumina powder, and to obtain required relative density with accuracy below 5% for higher pressures. It is planned to perform further investigations on repeatability of the densification process parameters and resulting density, and to introduce necessary corrections to the model.

4 Conclusions

From the performed theoretical analysis of the hot-pressing process under directly applied electrical current it can be concluded that differentiation between several densification mechanisms is crucial. In particular, plastic deformation, diffusion and creep of two kinds should be summed up when calculating the relative density of sintered specimen. Moreover, from the practical perspective, division of a sintering process to two stages and application of proper equations for each of them provided good estimation without necessity of distinguishing more stages.

Experimental results initially confirmed the correctness of the model. At lower pressure $P = 20$ MPa, difference between the experiment and the model was 7%, while at $P = 45$ MPa experimental results differed from the model by no more than ± 4%. Experiments also confirmed the theoretical prediction that at pressure $P = 45$ MPa the highest relative density can be obtained at 1350 °C after holding time $t = 5$ min, with no necessity of temperature increase or process prolongation. Additional experiments with alumina nanopowder with particles diameter 0.06 μm demonstrated possibility of full densification even at lower temperature of 1250 °C, in conformity with the theoretical model.

It is necessary, however, to perform more experiments in order to check the accuracy of the model. In particular, statistical analysis of the repeatability should be done, as well as other sintered materials should be considered. Nevertheless, from the practical perspective, the model can be applied for estimation of the process parameters to obtain the desired density of sintered alumina.

References

1. Abedi, M., Asadi, A., Vorotilo, S., Mukasyan, A.S.: A critical review on spark plasma sintering of copper and its alloys J. Mater. Sci. **56**(36), 19739–19766 (2021). https://doi.org/10.1007/s10853-021-06556-z
2. Munir, Z.A., Quach, D.V., Ohyanagi, M.: Electric field and current effects on sintering. In: Castro, R., van Benthem, K. (eds.) Sintering, pp. 137–158. Springer, Berlin, Heidelberg (2012). https://doi.org/10.1007/978-3-642-31009-6_7
3. Anselmi-Tamburini, U., Spinolo, G., Maglia, F., Tredici, I., Holland, T.B., Mukherjee, A.K.: Field Assisted Sintering Mechanisms. In: Castro, R., van Benthem, K. (eds.) Sintering, pp. 159–193. Springer, Berlin, Heidelberg (2012). https://doi.org/10.1007/978-3-642-31009-6_8
4. Swenson, O.F., Marinov, V.: Laser processing of direct-write nano-sized materials. In: Lawrence, J. (ed.), Advances in Laser Materials Processing. 2nd edn., pp. 571–594. Woodhead Publishing, Duxford, UK (2018). https://doi.org/10.1016/B978-0-08-101252-9.00019-4D.S
5. Yang, H., et al.: Unveiling exceptional sinterability of ultrafine α-Al_2O_3 nanopowders. J. Materiomics 7 4 837 844 https://doi.org/10.1016/j.jmat.2020.12.011
6. Sánchez-González, E., Miranda, P., Meléndez-Martínez, J.J., Guiberteau, F., Pajares, A.: Temperature dependence of mechanical properties of alumina up to the onset of creep. J. Eur. Ceram. Soc. **27**(11), 3345–3349 (2007). https://doi.org/10.1016/j.jeurceramsoc.2007.02.191
7. Sirota, V.V., Gevorkyan, E.S., Kovaleva, M.G., Ivanisenko, V.V.: Structure and properties of nanoporous ceramic Al2O3 obtained by isostatic pressing. Glass Ceramics (Engl. Transl. Steklo i Keramika) **69**(9–10) 342–345 (2013)
8. Wilkinson, D.S., Ashby, M.F.: Pressure sintering by power law creep. Acta Metall. **23**(11), 1277–1285 (1975). https://doi.org/10.1016/0001-6160(75)90136-4
9. Swinkels, F.B., Ashby, M.F.: A second report on sintering diagrams. Acta Metall. **29**(2), 259–281 (1981). https://doi.org/10.1016/0001-6160(81)90154-1

10. Wakai, F., Guillon, O., Okuma, G., Nishiyama, N.: Sintering forces acting among particles during sintering by grain-boundary/surface diffusion. J. Am. Ceram. Soc. **102**(2), 538–547 (2019)
11. De Jonghe, L.C., Rahaman, M.N.: Sintering of Ceramics. In: Somiya, S. et al. (eds.), Handbook of Advanced Ceramics, pp. 187–264, Elsevier, Amsterdam (2003)
12. Biesuz, M., Sglavo, V.M.: Electric forces effect on field-assisted sintering. J. Eur. Ceram. Soc. **40**(15), 6259–6265 (2020). https://doi.org/10.1016/j.jeurceramsoc.2020.06.069
13. Skorokhod, V.V., Uvarov, I.V., Ragulya, A.V.: Physico-chemical kinetics in the nanostructured systems. Akademperiodica, Kyiv (2001). (in Russian)
14. Raychenko, A.I.: Fundamentals of the process of sintering powders by passing electric current. Metalurgia, Moskva (1987). (in Russian)
15. Tamura, Y., Zapata-Solvas, E., Moshtaghioun, B.M., Gómez-García, D., Domínguez-Rodríguez, A.: Grain-boundary diffusion coefficient in α-Al2O3 from spark plasma sintering tests: evidence of collective motion of charge disconnections. Ceram. Int. **44**(15), 19044–19048 (2018). https://doi.org/10.1016/j.ceramint.2018.07.073
16. Maruyama, K.: Fundamental aspects of creep deformation and deformation mechanism map. In: Abe, F., Kern, T.U., Viswanathan, R. (eds.) Creep-Resistant Steels, pp. 265–278, Woodhead Publishing, Cambridge, UK (2008). https://doi.org/10.1533/9781845694012.2.265
17. Trapp, J., Semenov, A., Nöthe, M., Wallmersperger, T., Kieback, B.: Fundamental principles of spark plasma sintering of metals: part III – densification by plasticity and creep deformation. Powder Metall. **63**(5), 329–337 (2020). https://doi.org/10.1080/00325899.2020.1834748
18. Kislyi, P.S., Bodnaruk, N.N., Gorichok, N.I., Zaverukha, O.V., Kryl, Y.A, Kuzenkova, M.A.: Physicochemical fundamentals of the manufacturing of refractory superhard materials. Naukova Dumka, Kyiv (1986). (in Russian)
19. Gevorkyan, E., Lavrynenko, S., Rucki, M., Siemiątkowski, Z., Kislitsa, M.: Preparation of nanostructured materials by electrical sintering. In: Proceedings of the 7th International Conference on Mechanics and Materials in Design (M2D2017), pp. 663–666. Albufeira/Portugal (2017)
20. Gevorkyan, E., et al.: Sintered nanocomposites ZrO2-WC obtained with field assisted hot pressing. Compos. Struct. **259** 113443 (2021). https://doi.org/10.1016/j.compstruct.2020.113443

Fundamental Analysis of a Circular Metal Sawing Process

Dominik Brüggemann[1(✉)], Jens Kneifel[2], and Stefan Bracke[1]

[1] University of Wuppertal, Gaußstraße 20, 42119 Wuppertal, Germany
dbrueggemann@uni-wuppertal.de
[2] FGW Remscheid, Papenberger Str. 49, 42859 Remscheid, Germany

Abstract. The most appropriate separation technique for the processing of solid metal parts with large dimensions is sawing. The cutting tools used in this machining process are exposed to very high mechanical and thermal loads, yet the highest precision, product quality and process stability must be guaranteed. With regard to process optimisation, the prediction of tool failure and the estimation of the remaining useful life is an important goal. In this paper, a first approach is presented to work towards the development of a degradation model for circular saw blades based on monitored process parameters. Such a degradation model could then be used to derive the key objectives presented above. The basis of this analysis is the recording of the sensor signals current, voltage, vibration and sound with a high sampling rate, whereby in the present work the focus will initially be on the first two signals mentioned - current and voltage. These signals were analysed in such a way that key indicators could be derived. These key indicators were then used to carry out initial analyses, which are intended both to increase the understanding of the process and to form the basis for future analyses.

Keywords: Circular sawing process · Key indicator system · Multivariate process analysis · Process data acquisition · Process data processing

1 Introduction

Machining (metal cutting) is the collective term for a group of subtractive manufacturing processes that give workpieces a specific geometric shape by mechanically removing excess material from raw parts in the form of chips. The most important machining processes are turning, drilling, milling, sawing and grinding. According to the German Industrial Standard DIN 8589-0 [1], cutting belongs to the main group of machining and is the most important group in terms of industrial significance.

The most appropriate separation technique for the processing of solid metal parts with large dimensions such as bars, shafts, discs, blocks, tubes or rings is sawing. The cutting tools (industrial saws) are designed as an endless band, as a saw blade of a reciprocating saw or as a circular saw blade and typically have saw teeth with cutting edges at the rim. The separation itself is caused by a relative motion between the sawing tool and the workpiece such that a narrow kerf is cut. Each individual saw tooth scrapes

B. Gapiński et al. (Eds.): MANUFACTURING 2022, LNME, pp. 124–137, 2022.
https://doi.org/10.1007/978-3-031-00805-4_11

off a thin chip, which is transported out of the workpiece in a saw tooth gap. At the same time, mechanical forces and high temperatures affect the saw teeth, such that wear occurs. The mechanical and thermal loads depend on various parameters, such as the material of the sawing tool, the material of the workpiece, the cutting speed, the feed rate, the cutting pressure and the cutting edge geometry. The mechanical wear accounts for the majority of wear at low cutting speeds and when machining easily machinable materials like unalloyed steels. In contrast, at high cutting speeds and materials that are more difficult to cut (e.g. stainless steels), thermal wear like oxidation and diffusion is caused and is responsible for the main part of the wear. During the machining process several additional wear mechanisms occur in addition to the previously mentioned. These include abrasion due to friction between chip and sawing tool and adhesion. Most of the wear appears at the flank and the face of the tooth (see Fig. 1). Flank wear is generally caused by high temperatures due to friction. Crater wear occurs at the face of the cutting tool due to the removal of chips [2].

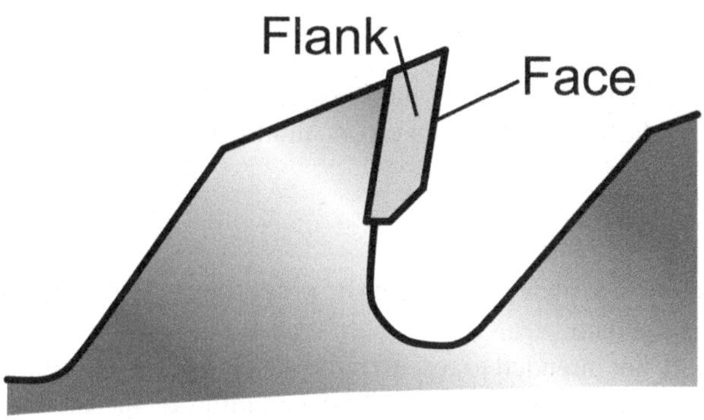

Fig. 1. Schematic representation of a tooth of a circular saw blade.

In addition to this gradual wear during regular operation, the failure modes of blade breakage and tooth breakage occur in machining with circular saw blades. While blade breakage is caused by incorrect positioning during the first cut or the use of blunt circular saw blades, tooth breakage is due to the choice of an incorrect tooth pitch (i.e. tooth pitch is unsuitable for the material cross-section) [3].

Due to the aforementioned effects, the condition of the saw blade changes over the duration of its use. Monitoring this change is a crucial factor in ensuring the quality of the tool application and avoiding costs for unnecessary damage or destruction of the tool. Such condition monitoring systems are usually referred to as Tool Condition Monitoring Systems (or TCMS for short). TCMS create the possibility to economically and ecologically optimize processes in metal cutting. This monitoring and evaluation of the transmitted data is an essential part of the idea of Industry 4.0 [4]. TCMS are able to monitor the state of the tool by analysing sensor data and extracting information in form of key indicators from them [5].

With the machines used, it is possible to measure parameters such as current and voltage with non-invasive sensors, which have no influence on the running process [6]. The sensors are used to monitor the individual cuts of the saw blade. Based on the changes in the measured values, it is believed that it is possible to monitor the wear of tools. In many modern machines, sensors that monitor the energy supply of machines often do not have to be retrofitted, but are already installed by the manufacturer of the machine [7]. In the course of the paper, the sensors used for these analyses will be examined in detail again.

In contrast to milling, drilling or turning, the wear behaviour of circular saw blades has not yet been researched as thoroughly. In order to determine the wear, process data must be recorded continuously when the saw is in use. The optimal choice of process data to be observed for quantifying wear varies from case to case. Common methods include the use of force sensors, torque sensors, power sensors, temperature sensors, vibration sensors and acoustic emission sensors [7].

This paper investigates whether wear behaviour of circular saw blades can be detected by collecting motor current data and motor voltage data. According to Ammouri, and Hamade [8], the current increases when the wear on the tool increases, because the cutting force increases. In milling processes, studies on the use of motor current to detect wear have proven to be a successful and simple method that can significantly reduce the complexity of data collection.

While the use of multiple sensors increases the accuracy of wear detection, it also increases the costs, the susceptibility to errors and the maintenance effort [5, 9, 10].

2 Goal of the Study

The goal of this study is to understand the basic properties of the process and to recognise basic structures in the data.

It is deliberately not intended to research the deeper interrelationships of such recognised data structures, as the fundamental understanding of the process at hand that is being sought is an absolute necessity for such studies.

For this reason, the section "Fundamentals" first explains which process parameters are relevant for the subsequent analysis of the process under consideration. Furthermore, properties of the process are highlighted and defined there, which is a prerequisite for the analysis of the process. In the further course of the section, the sensors used are first considered before the used data set is presented in the last subsection.

Using this data set as an example, the fourth section then presents the processing of a data set and the development and application of a key indicator system.

In the fifth section, a first approach, which can serve as a basis for the development of a degradation model, is presented. Here, the entire train of thought is deliberately traced in order to show the underlying discovery process.

In the last section, besides the summary, an outlook is given, which presents possible future studies that can build upon this paper.

3 Fundamentals

3.1 Parameters and Notations

The process parameters, variables and geometry characteristics of a sawing process are summarized and defined in the following table below.

Table 1. Parameter designations of circular sawing processes.

Parameter name	Formula notation	Unit
Saw blade diameter	d_S	[mm]
Saw blade radius	r_S	[mm]
Saw blade circumference	U_S	[mm]
Number of teeth	Z	[-]
Workpiece height	h_W	[mm]
Workpiece width	b_W	[mm]
Rotational speed of saw blade	n	[1/min]
Angular velocity of saw blade	ω	[1/s]
Cutting speed of saw blade	v_C	[m/min]
Rotation time of saw blade	T	[s]
Engagement time of one tooth	t_C	[s]
Tooth feed rate	f_c	[mm/tooth]
Total feed rate	v_f	[mm/min]
Depth of cut	h	[mm]
Width of cut	b	[mm]
Length of cutting edge	l_{ce}	[mm]
Duration of cut	D_C	[s]

The parameters are separated into four categories. The first category describes properties of the saw blade. The second category describes properties of the workpiece. The third category contains setting parameters of the sawing process. The fourth category contains terms which are used to describe the current state of a sawing process.

In the data set, which will be examined in more detail later, the parameters d_s (thus also r_s and U_s), Z, h_W and b_W are uniform, i.e. they are the same for all processes included. In general, however, this statement does not hold.

For the cutting speed of the saw blade v_c (and thus n, ω, T and t_c) the dataset contains numerous processes with deviating values. The same holds for the feed rate v_f (and thus f_c).

The depth of cut indicates how far the saw blade has penetrated the workpiece in the vertical direction. The cutting width indicates how far the saw blade has penetrated the workpiece in the horizontal direction. The length of the cutting edge depends on the saw blade radius r_S as well as the cutting depth and is limited by the dimensions of the workpiece. Some of the notations defined in Table 1 are visualised in the schematic diagram in Fig. 2.

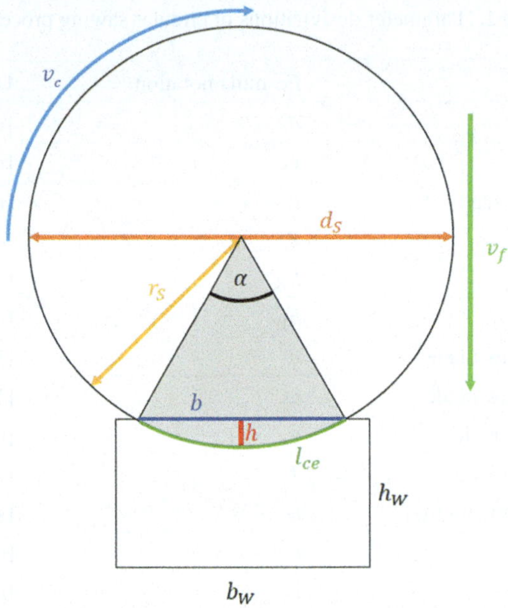

Fig. 2. Schematic representation of a circular sawing process with rectangular workpiece.

3.2 Course of the Sawing Process

In order to be able to analyse the course of a sawing process in a differentiated manner, the course was divided into 5 different phases, which relate to the position of the saw blade to the workpiece. The phases are separated from each other by the separation points 1–4.

Circular sawing of metal in industry usually involves the workpiece being clamped firmly or fixed in a vice while the saw is guided slowly through the material. The workpieces considered in this paper all have a rectangular cross-section. The machine operates at relatively low speeds (usually between 70 and 170 m/min cutting speed) and slow linear feed (usually between 0.04 and 0.10 mm per tooth).

The phases listed below are additionally visualised in Fig. 3 using the length of the cutting edge as the cutting depth progresses. In addition, the positions of the saw blade to the workpiece corresponding to the phases are shown schematically in the upper part of the graphic.

The 5 phases are as follows:

- Phase 1: Idling (no load). The circular saw blade has no contact with the workpiece, but is already rotating at the constant cutting speed. The sensors therefore detect the basic load resulting from the rotary movement.
- Phase 2: Increase of the length of the cutting edge. The cutting tool starts penetrating the workpiece until maximum engagement is reached. The maximum engagement is governed by both the dimensions of the saw blade and the dimensions of the workpiece. The increase in the length of the cutting edge is not linear.

- Phase 3: Constant behaviour of the length of the cut edge. This phase of the process represents the longest part of a cut in terms of time. The phase lasts until the saw blade starts to exit at the bottom end of the workpiece.
- Phase 4: The length of the cutting edge decreases as soon as the saw blade exits the lower end of the workpiece. This decrease, like the increase in phase 2, is not linear.
- Phase 5: Idling (no load). The cutting process is completed and similar to stage 1 the tool is not in contact with the workpiece.

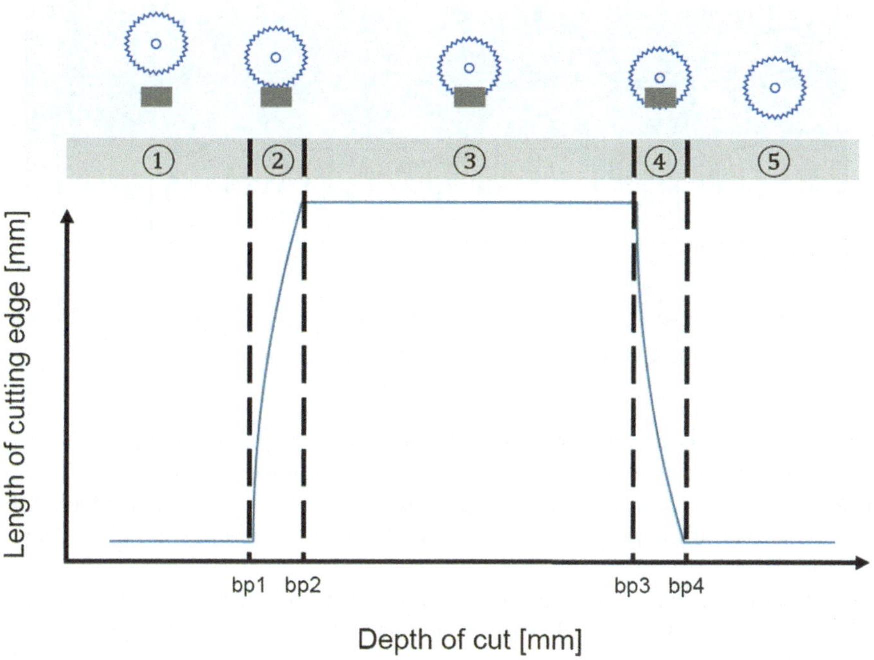

Fig. 3. Exemplary sequence of a cut with change of the characteristic tooling parameters.

3.3 Sensors, Measurement Technology and Production Data Acquisition

For data collection, only non-invasive methods/sensors were used, as previously mentioned. In particular, custom milled parts were created to attach two triaxial accelerometers (ICP® from PCB, model number 356A02) on existing boreholes that are usually used for assembly. One accelerometer is mounted at the level of the shaft on which the circular saw blade is attached and the other on one side of the vice, right next to where the saw advances through the workpiece. The sound pressure is picked up by a prepolarized free-field microphone and an amplifier (both from Brüel & Kjær, type 4189 and 2671) located outside of the main enclosure, to ensure that the unit receives a signal from a similar distance as the machine operators are used to hearing the sounds.. The electric current and amperage is measured with Rogowski coils and signal converters located at the main power source of the machine.

The data acquisition runs on a NI PXI system with synchronized multi I/O-modules and a custom software to monitor the high frequency, high data rate signals, and detect active sawing processes (cf. Fig. 4).

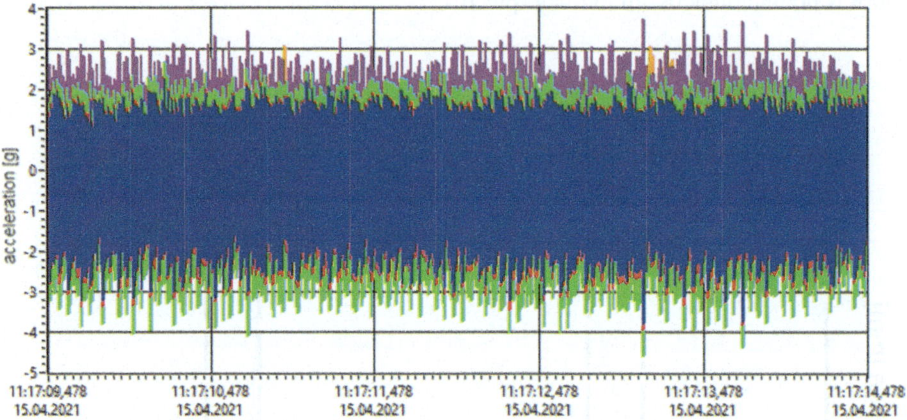

Fig. 4. Example of measurement from IEPE accelerometers.

In the data set discussed in the paper, the individual process recordings were started manually by the operator of the machine. Since this unfortunately led to very inconsistent and, above all, incomplete recordings, a light barrier has now been installed which detects the start and end of the process to be measured and starts or stops the data recording accordingly.

The last component of the data set is handwritten protocols in which the machine operators note down the selected process parameters (cutting speed, feed, material, saw blade number) and observations at regular but widely spaced intervals (usually 2 h). Due to the very large intervals, the logs play only a minor role in the analysis of the data set. They are mainly used to make general cut-offs of the data set, e.g. when changing the saw blade, as only the cuts of one saw blade are to be analysed.

4 Analysis of the Timeseries

4.1 Data Set

In the dataset under consideration, 76 complete sawing processes are documented whereas 5 incomplete processes had to be removed from the dataset. Here, 9 signals are recorded at a time. Two signals, the current [A] and the voltage [V] are recorded with a sampling rate of 1000 data points per second. The other seven signals come from the two piezo sensors (3 signals each) and a microphone. They are each recorded with a sampling rate of 20,000 data points per second.

The length of a single sawing process in the data set ranges from 154.6 s to 259.1 s. The average length is 193.1 s.

The focus of the following initial evaluation of a dataset shall be on the signals of the current and the voltage as the analysis of the datasets is still in an initial stage and the signals shall be examined in small portions. In future evaluations, it is also planned to investigate larger portions of signals simultaneously and their dependencies and influences on each other.

4.2 Separating the Processes

In order to be able to look at the processes in a differentiated way, it is necessary to separate the processes into the process phases explained in Sect. 3.2. For this purpose, a detection mechanism for the separating points was developed, which is presented in the following.

First, the signal is smoothed, as it is subject to strong scattering, which can lead to distortions during separation point detection. For smoothing, a moving average with a window width of 10,000 values is performed, which corresponds to an average of the signal over 10 s. The smoothed signal of a process from the data set can be seen in Fig. 5. Subsequently, a gradient function of the smoothed signal is formed. For this, the average slope is determined over a window width of 2 s. This approximation of the local slope is also shown in Fig. 5.

To determine the separation points, the two main changes in the current signal are first determined (increase in the curve in the first half due to the saw penetrating the workpiece [phase 2], decrease in the curve in the second half due to the saw leaving the workpiece [phase 4]). For this purpose, the time of the maximum value of the slope within the first half of the process as well as the time of the minimum value of the slope within the second half of the process are determined. In the case shown in Fig. 5, these extrema occur at approximately 18.1 s and 150.7 s, respectively. Starting from these two points in time, the latest point in time before and the earliest point in time after those extrema are determined at which the slope assumes a value in a small range around zero. In the evaluations of this paper, a limit value of 0.2 was chosen for this. These points in time correspond to the separation points sought, since each of these separation points is adjacent to a phase which is characterised by a largely constant course and thus a low and thus near-zero gradient, as can also be seen in the principle sketch in Fig. 5. The separation points are marked by vertical black lines in the figure.

The basic idea for this separation point detection mechanism was developed specifically for the process structure at hand. The parameters were adapted for a small group of randomly selected processes and then verified for the entire present data set by visual verification. Although it would have been possible to determine the separation points by hand with the relatively small data set at hand and it would have been similarly time-consuming, it seemed important to us to automate this process. On the one hand, there are other data sets for the same process, and on the other hand, the detection of the separation points should be repeatable and consistent, which is given by a value-based mechanism. The consistency of the separation point detection will be addressed again in this paper within the context of the key indicator analysis.

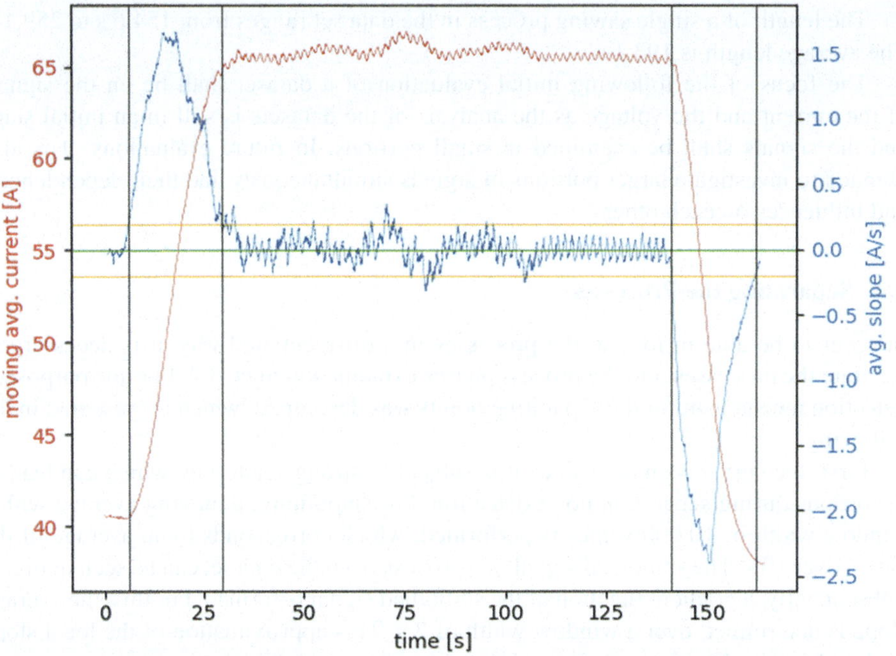

Fig. 5. Smoothed signal (red), average slope (blue) and separation points (black) of a process of the considered data set. The separation points were determined using the average slope and the selected limit values of 0.2 (orange).

4.3 Creating a System of Key Indicators

As already mentioned, a system of key indicators was created on the basis of the previously identified separation points. The aim of such a key indicator system is to reduce a time series to a few key indicators while retaining the information content contained in the time series. Indicator systems such as this one, which are used in the analysis of technically complex systems and processes, have delivered good results in the past [11, 12].

The first recorded key indicators are the times of the separation points, measured in seconds from the beginning of the data recording of the process under consideration. In addition, the time of the separation points is also recorded in relation to the total length of the data recording.

The lengths of the various identified phases measured in seconds, are also recorded in order to be able to carry out a verification of the separation point detection.

Furthermore, characteristics of the signals within the identified phases are determined. These include the mean intensity, the mean slope or the dispersion of current, voltage and their product, power.

A total of 85 key indicators are thus recorded per process, some of which will be looked at in more detail below.

As mentioned, only some initial analyses are carried out in this paper. However, the key indicator system can also be used in further, more sophisticated analyses. For example, an analysis of the data by machine learning methods can be based on such an indicator system.

4.4 Verification of the Separation Point Detection

To verify whether the developed separation point detection yields the desired consistent and repeatable results, we examined the lengths determined for the third phase of the processes.

The third phase was chosen because it was the only phase that was entirely available in all processes. The other phases were not entirely included in a few cases due to manual starting and stopping of the data recording.

As explained before, the length of the third phase is determined by the time of the separation points 2 and 3.

The determined length of the third phase for the 76 processes considered in the data set can be seen in Fig. 6.

As can be seen, there are clearly some plateaus in the determined length of the third phases. The duration of the phases remains largely constant over the length of these plateaus and deviates only by small differences.

The occurrence of these plateaus can be explained by the different feed rates selected by the machine operator. The pattern that is revealed is consistent with the handwritten protocols in which the feed rates are noted as well as with the known procedure of the machine operators, who leave the feed rate unchanged until problems occur in the process.

The figure shows that the first 18 processes have an almost identical length of the third phase, which lasts an average of 110.4 s.

Then there are 34 processes whose third phase lasts a little longer. The average length here is 113.6 s. This slightly increased duration can be explained by a slightly reduced feed rate, which means that the saw blade needs a little longer to pass through the workpiece.

Another plateau is formed by 9 further processes with an average length of the third phase of 148.3 s.

The last plateau shows 15 processes whose third phase takes an average of 130.1 s. Based on these clearly recognisable plateaus, it can be assumed that the separation point detection works well and consistently.

5 Approach for the Development of a Degradation Model

In an example, a first approach for a degradation model was derived based on the key indicators described in Sect. 4. Within this example the usage of one circular sawblade was considered from first cut to failure. The considered example consists of 76 cuts where the dimensions and the alloy of the workpiece were constant. Therefore, the influence of cutting different kind of workpieces was neglected.

The basic idea of this first approach was to relate the degradation of the circular saw blades to the power consumption during the cut. In this example, the consideration of

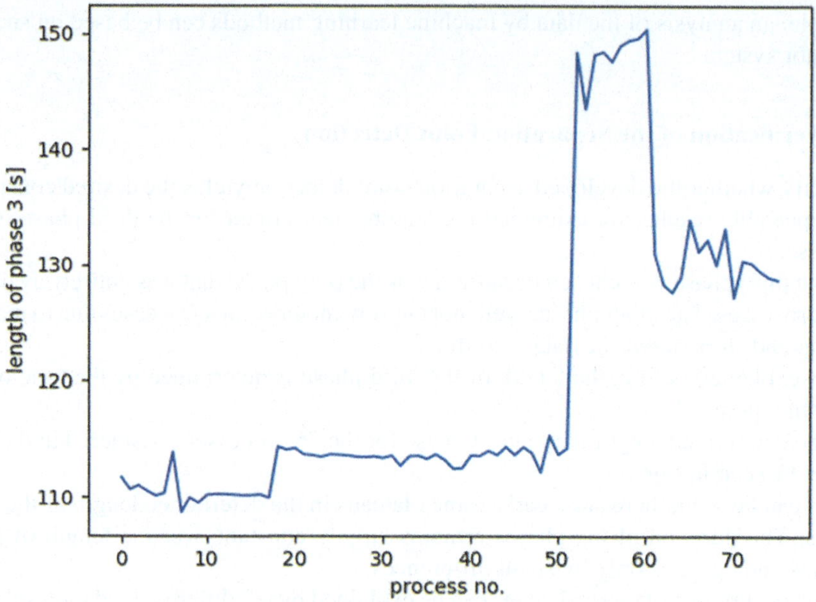

Fig. 6. Length of the third phase of the processes.

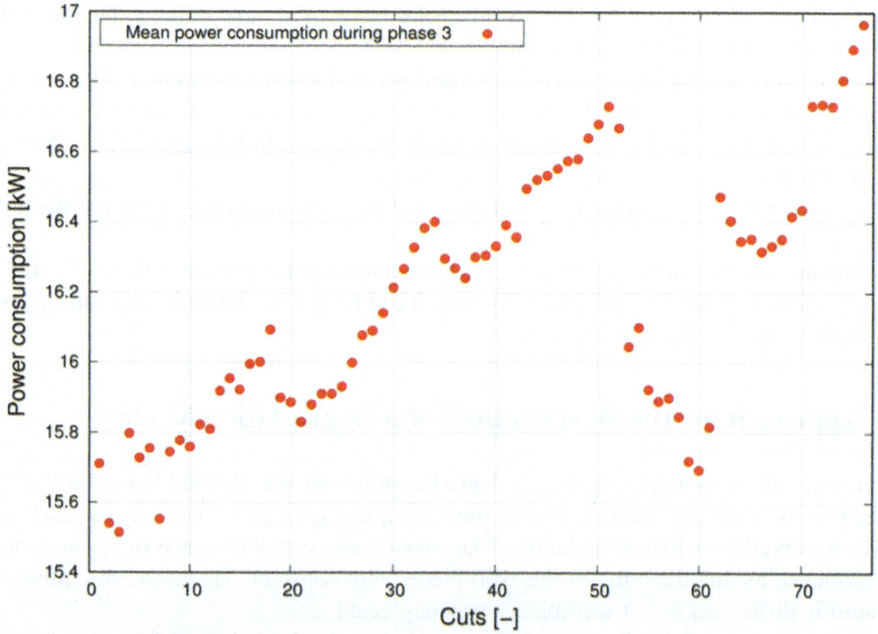

Fig. 7. Power consumption during phase 3 of each cut.

the power consumption was limited to the third phase of the cutting process. In Fig. 7 the mean power consumption during phase 3 of each cut is shown.

It can be seen, that the mean power consumption in phase 3 is increasing over the course of the cuts until a decrease after 52 cuts can be observed. This decrease of power consumption can be related to a variation of the process parameters. It was assumed, that the decrease of the mean power consumption in phase 3 is caused by a decrease of the linear feed. This assumption is consistent with the observations presented in Sect. 4.4. A decrease of the linear feed causes an increase of the duration of a cut. It is therefore reasonable to consider the process time in addition to the mean power consumption. In Fig. 8 the energy consumption during phase 3 of each cut is plotted.

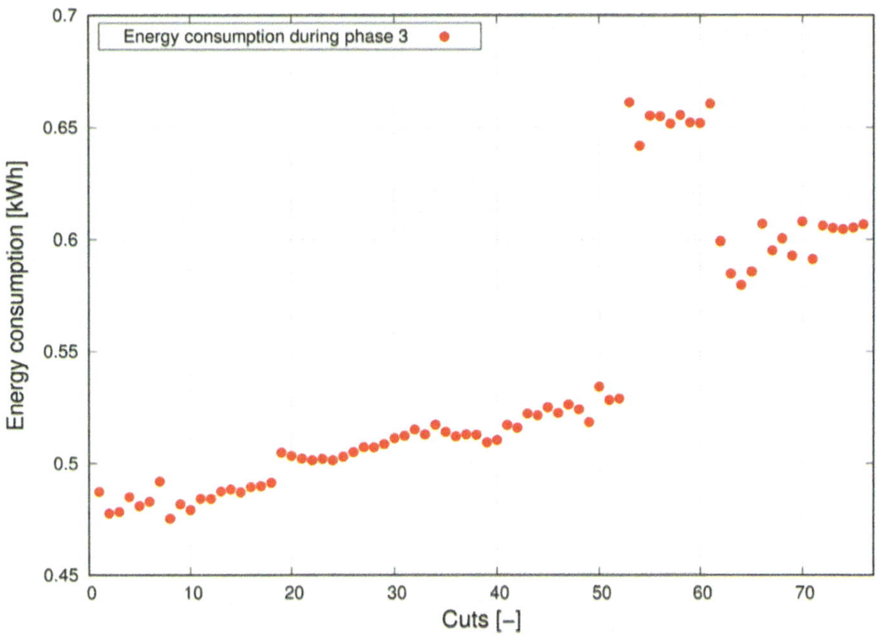

Fig. 8. Energy consumption during phase 3 of each cut.

Figure 8 shows that the energy consumption increases drastically after 52 cuts, suggesting the variation of another process parameter: the cutting speed. Increasing the cutting speed requires an increased engine speed and therefore a higher energy demand. This effect can be taken into account if the energy consumption in idle mode (phase 1 and phase 5) is considered. The energy consumption in idle mode was subtracted from the energy consumption in phase 3. The resulting cutting energy consumption is shown in Fig. 9.

It can be observed, that the cutting energy consumption is on a constant level for the first 30 cuts where after an increase of the cutting energy consumption sets in until failure occurs.

The development of a general degradation model for circular saw blades based on monitored process parameters requires comparability of usage. In the presented example a method to consider the influence of rather unmonitored process parameters like the linear feed and the cutting speed was shown (as mentioned before, the process parameters used are only recorded approx. every 2 h). As a result, the degradation of circular saw blades can be related to the increase of the cutting energy consumption over the course of the usage phase.

In further studies, it needs to be analysed, how the influence of cutting work pieces with varying dimensions and alloys can be considered for the development of a general degradation model.

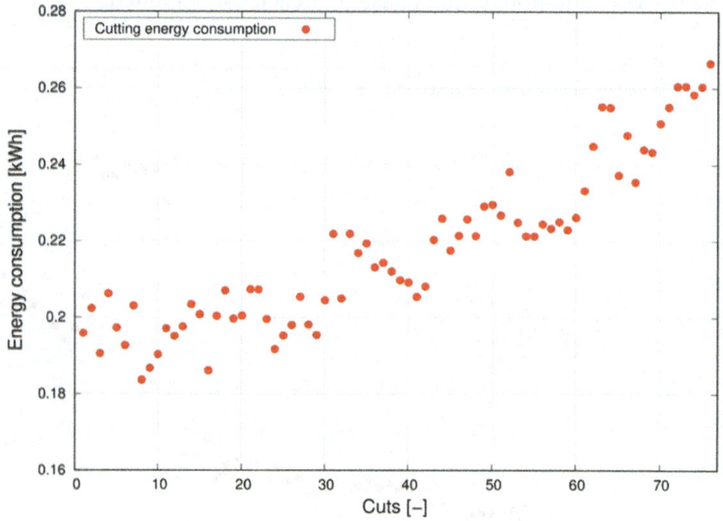

Fig. 9. Cutting energy of each cut.

6 Summary and Outlook

The present research work deals with the analysis of sawing processes with the focus on sensor signal acquisition and processing of the electric current data. After a process separation for each cut, time-based features as well as current values (e.g. amperage and voltage) can be extracted to perform a degradation analysis for the tool. The results show that with increasing tool wear, there is an increase of electrical current and energy consumption as well. This approach provides good results, which can be used as a basis for future in-depth studies on the degradation behaviour of the tool, especially if the cutting speed, the process time and the idle phase influence are also considered. Future studies will therefore focus on these but needs to be verified in future studies.

Acknowledgements. The authors thank Christoph Rosebrock, Max Radetzky, Tom Stürwold and Pit Fiur for the contributions. Furthermore, the authors thank the project sponsor ERDF (European Regional Development Fund) and the Ministry of Economic Affairs, Innovation, Digitalization and Energy of the State of North Rhine-Westphalia. The project organization is supported by PTJ (Project Management Jülich).

EUROPÄISCHE UNION
Investition in unsere Zukunft
Europäischer Fonds
für regionale Entwicklung
EFRE.NRW
Investitionen in Wachstum
und Beschäftigung

References

1. DIN 8589-0, Manufacturing processes chip removal - part 0: General; classification, subdivision, terms and definitions. Beuth Verlag, Berlin (2003). Qreg
2. Tönshoff, H.K., Denkena, B.: Basics of Cutting and Abrasive Processes. Springer, Heidelberg (2013)
3. Tschätsch, H.: Applied Machining Technology. Springer, Heidelberg (2009)
4. Kuntoglu, M., et al.: A review of indirect tool condition monitoring systems and decision-making methods in turning: critical analysis and trends. MDPI (2020)
5. Zhou, Y., Sun, W.: Tool wear condition monitoring in milling process based on current sensors. IEEE Access **8**, 95491–95502 (2020). https://doi.org/10.1109/ACCESS.2020.2995586
6. Sevilla-Camacho, P.Y., Robles-Ocampo, J.B., Muñiz-Soria, J., Lee-Orantes, F.: Tool failure detection method for high-speed milling using vibration signal and reconfigurable bandpass digital filtering. Int. J. Adv. Manuf. Technol. **81**(5–8), 1187–1194 (2015). https://doi.org/10.1007/s00170-015-7302-0
7. Tönshoff, H.K., Insaki, I.: Sensors in Manufacturing. Wley-VCH Verlag GmbH (2001)
8. Ammouri, A.H., Hamade, R.F.: Current rise criterion: a process-independent method for tool-condition monitoring and prognostics. Int. J. Adv. Manuf. Technol. **72**(1–4), 509–519 (2014). https://doi.org/10.1007/s00170-014-5679-9
9. Stavropoulos, P., Papacharalampopoulos, A., Vasiliadis, E., Chryssolouris, G.: Tool wear predictability estimation in milling based on multi-sensorial data. Int. J. Adv. Manuf. Technol. **82**(1–4), 509–521 (2015). https://doi.org/10.1007/s00170-015-7317-6
10. Jahromi, A., Er, M., Li, X., Lim, B.: Sequential fuzzy clustering based dynamic fuzzy neural network for fault diagnosis and prognosis. Neurocomputing **196**, 31–41 (2016). https://doi.org/10.1016/j.neucom.2016.02.036
11. Hinz, M., Brüggemann, D., Bracke, S.: On the application of machine learning techniques in condition monitoring systems of complex machines. In: 14th Probabilistic Safety Assessment and Management Conference - PSAM, Los Angeles (USA)
12. Hinz, M., Brüggemann, D., Bracke, S.: Framework for customized, machine learning driven condition monitoring system for manufacturing. In: 25th International Conference on Production Research - ICPR, Chicago (USA)

Advantages of Automatic CAM Programming in Industrial Practice – A Case Study

Maciej Kowalski and Przemysław Zawadzki(✉)

Faculty of Mechanical Engineering, Poznan University of Technology, Piotrowo 3, 60-965 Poznań, Poland
przemyslaw.zawadzki@put.poznan.pl

Abstract. The paper presents the implementation of automatic CAM programming for the manufacture of special production tooling. The programming procedure, called ACPUT (Automatic CAM Programming Using Machining Templates), includes the development of special machining templates in the CAM program, supported by the acquisition of technological knowledge in a specially prepared database. These templates are dedicated to a given group of parts, characterized by the similarity of their geometric features. ACPUT makes it possible to shorten the time required to develop a machining program, thereby having a positive impact on the total cost of tooling production. The aim of the work was to present the advantages of the use of automatic CAM programming compared to the traditional approach used by technicians/programmers with different levels of experience (expert and beginner). The tests were carried out on CAM programming process for welding tooling.

Keywords: CAM programming automation · KBE - Knowledge-Based Engineering · Machining templates

1 Introduction

To meet the expectations of the mass personalization of products, it is not enough to simply invest in modern machinery. It is necessary to develop digital systems that allow for the comprehensive use of various data gathered in product designs and its production technologies [1–4]. The implementation of such systems requires the appropriate organization of departments responsible for the preparation of technical documentation (construction and technological). The use of CAx systems in technical departments has been a standard for many years. Thanks to this, designers and technicians can use tools that make it possible to use the knowledge gained in previously implemented projects (like KBE - Knowledge-Based Engineering). Such smart solutions are used primarily in companies dealing in the mass production of technologically similar, multi-variant products [5, 6]. An example of this can be found in the automotive sector, where similar components having the same purpose may differ on account of different vehicle variants. The same is true in other industries, such as in the manufacture of household appliances, sanitary or electronic products.

B. Gapiński et al. (Eds.): MANUFACTURING 2022, LNME, pp. 138–148, 2022.
https://doi.org/10.1007/978-3-031-00805-4_12

It should be emphasized that the plan of the technological process, next to the engineering design, is undoubtedly the most important part of the technical preparation of the production process. It concerns the planning of elementary activities related to the production of semi-finished products, final products, and tools necessary for their production. In other words, it is about "how to realize what the designers have designed". An important step of the technological process in modern production systems is the stage of programming of CNC machines, which is performed in CAM software. The process of machine programming in the CAM program determines the basic parameters of the machine, such as the number of controlled axes, dimensions of the machining table, movements along individual axes, etc. Moreover, the programming process involves determining the geometry of the stock and of the workpiece. In individual machining operations, the CAM programmer determines the specific geometric elements of the CAD model for processing (planes, edges, points), the machining tool type and features, the machining strategy (the machining tool path), and the specific parameters of the operation (speeds and the input/output paths of the machining tool relative to the workpiece).

Basis on that the CAM programming process is characterized by its repeatability. This means that certain activities are performed in every case, no matter what the specific project is. Since the process is repeatable, it is possible to automate it - at least in theory, because although there are various ways of automating work in CAM, the use of such solutions is not very popular in practice. Based on own observations and experience, the authors conclude that there are two main reasons for such a state of affairs. The first is caused by the level of education and experience of CAM programmers, and the second is related to the characteristics of the enterprise, the products themselves, and their manufacturing processes (high diversity and variability).

High efficiency of CAM programming or rather the quality of programs prepared in the CAM system (according to the principle – "do it right at the first time") requires high skills of using CAM software and knowledge about manufacturing process (e.g. milling, turning, etc.). Engineers at the basic level of skills are not able to prepare advanced, dedicated tools, because is not easy. Additionally, some of these tools require computer programming skills in the particular programming languages to operate. The preparation of dedicated solutions to aided engineering work in CAM software often requires the involvement of interdisciplinary teams.

The second reason for limiting automation in CAM is the characteristics of the processes carried out in the enterprise, or rather the level of variability of the manufactured products. Automation of CAM programming will be reasonable only in selected cases, where there is high variability in production and relatively its high similarity. In the case of single, unique work, CAM automation tools seem completely unnecessary. Each project is different, it should be analyzed separately and unique solutions should be selected for it. Moreover, such a program will probably be used only once (although the knowledge from such a project should be gathered despite this). Tool for automation of CAM programming will be used wherever many products are produced similar to each other (multi-variants), regardless of the scale of this production (i.e. small, medium, or large series).

To meet these needs, the authors developed their own methodology for preparing an automatic system of CAM programming, called ACPUT (Automatic CAM Programming Using Machining Templates). Its basic assumptions and some tools were described in detail in earlier works [7–10]. The main aim of this paper is to present the advantages of using ACPUT in industrial practice, presenting the results of automation of CAM programming on a selected example of special production tooling – matrix used in the welding process.

2 Research Methodology

2.1 Feature Recognition

The most popular tools for automating routine, repetitive tasks in a CAM system are based on recognizing the characteristic features of a CAD model (so-called FR – feature recognition technology). FR allows the assignment of appropriate machining cycles to specific recognized geometry shapes [11]. In practice, these tools can actually speed up the work on program preparation but are only effective for models with relatively simple shapes. In the case of advanced surface models (such as the dies or matrixes with shaping cavities described in this article), the proportion of automatic recognition of geometric features decreases, and the "manual" work of the technicians increases to a large extent. These tools are therefore a hybrid solution that works well in a single case. Therefore, for a group of families of similar products (several - several dozen items), their use does not significantly affect the effectiveness of the CAM programming process.

Although the FR technology have been known for three decades, they are still the subject of various works [12]. The authors focus on the development of various algorithms that can more accurately analyze 3D models and more effectively indicate possible technological operations [13, 14]. In turn, Zhou et al. [15] presented the FR method (supported by deep learning) for the selection of cutting tools, increasing the effectiveness and efficiency of this task. Another example is the use of FR for spot welding recognition. In turn, Chee Fai Tan et al. [16] described the methodology for recognizing the features of the CAD model, based on which appropriate operations in the CAM program are selected. Similar considerations were described in papers [17, 18], presenting different approaches to the data exchange between the CAD and CAM programs, as well as to recognition of the features of the objects processed, using universal STEP files format.

It should be noted that these works do not deal with the issue of the automation of CAM programming in a general perspective, which can actually improve the efficiency of this process, but only develops algorithms for searching for specific geometric features of 3D models.

2.2 ACPUT Procedure

Based on KBE approach a procedure for creating machining templates in the CAM system was developed. ACPUT in conjunction with a machining knowledge base, makes it possible to automate the operations in developing machining program for CNC machine [9, 10]. The ACPUT method is pointed to a specified product group (family of multi-variants products), and the rules it creates are effective and efficient in the whole group. The templates can be developed using any integrated CAD/CAM class software that allows the parametric construction of CAD models and programming in a scripting language (e.g. VBA – Visual Basic for Applications). Since machine programming in CAM systems is a time-consuming and high-cost process (it requires the involvement of highly qualified engineering staff), the aim of the ACPUT was to reduce time and minimize human participation in the programming of CNC machines (in the manufacture of special production equipment).

ACPUT assumes to develop a special machining template that represents all the technological operations possible for a given group of products. Data and information needed to prepare such a template are stored in the special knowledge base. Knowledge acquired mainly from specialists in a given field is accumulated and written formally, and in a way understood by the computer program. To ensure the correct operation of the machining template, and thus the automation of the preparation process in the CAM environment, firstly3D models for the tooling must be properly described in the CAD program (i.e. categorization of features and their assignation to the model). Based on this description, the template later automatically selects the appropriate features for the given part.

It is possible to look at ACPUT as a procedure, that includes the following steps (Fig. 1):

1) Analysis of 3D CAD models of a given group of tooling parts.
2) Preparation of the technological knowledge base.
3) Defining the geometry necessary to build the machining template and publishing it.
4) Preparation of machining template.
5) Preparation of machining program for each part of the given group of tooling parts (with simulation in CAM program).
6) Preparation of NC program.

Execution of the ACPUT procedure requires different skills in the field of manufacturing knowledge and CAx systems operation. It is therefore assumed that Step 1 and 2 are realized by technicians with large of experience. Step 3 can be carried out even by novice CAM programmer; however, Step 4 requires skills at expert CAM programmer level. Last two steps can be realized also by novice CAM programmer.

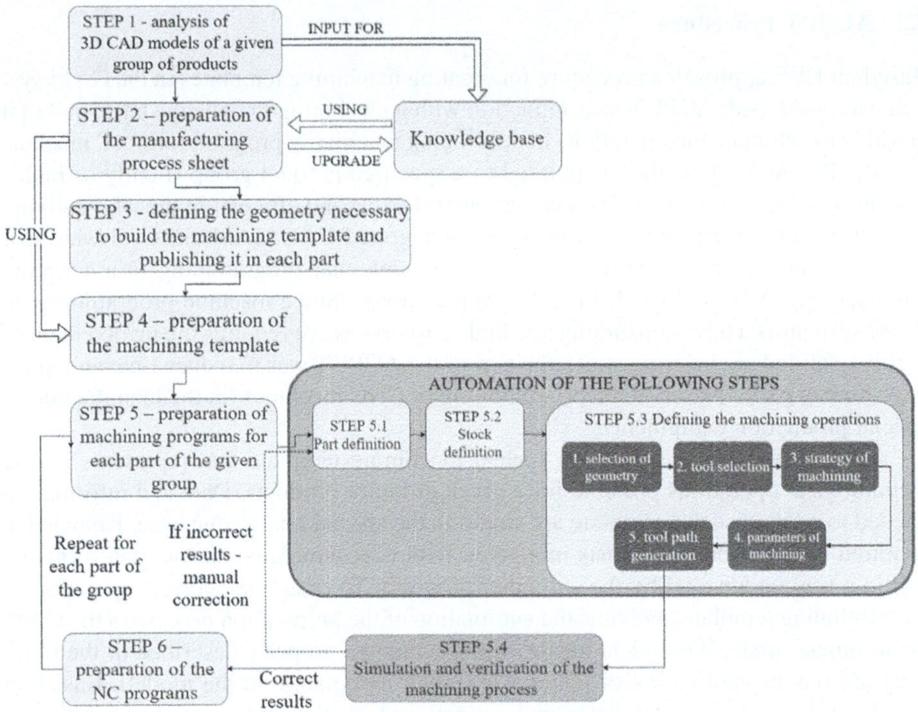

Fig. 1. The ACPUT procedure [10]

3 Case Study – CAM Programming for Welding Tooling Machining

The example, considered in this paper, is related to the machining of tooling elements for welding processes. Special production tooling in this case consists of 10 separate parts (Fig. 2) and manufactured by milling on a CNC machine. Acting in accordance with the assumptions of ETO (the tooling model was provided by the customer), it was assumed that in preparing programs it was not possible to interfere with the geometry of the 3D models of the tools, and the files were saved in the universal "stp" format.

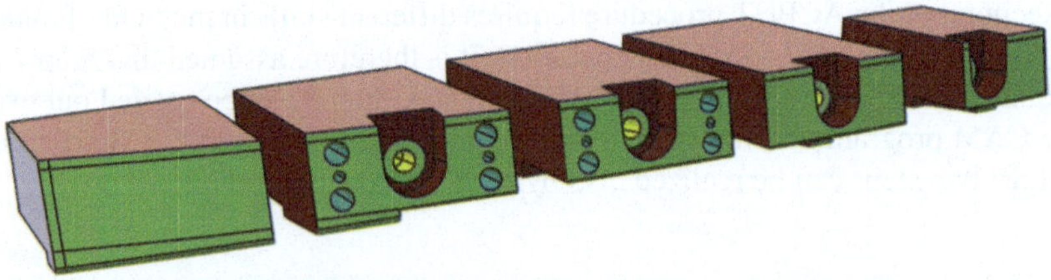

Fig. 2. Part of group of special tooling for welding

In accordance with the ACPUT procedure, ten different operations were identified and described, for which the detailed course of the machining process was recorded in the knowledge base. It consisted of (Fig. 3):

- machining of an oval pocket (here for the so-called suction cup),
- circular milling of the surface for the suction cup,
- drilling holes (various types), and
- roughing and finishing of the working surface of the tool (green in Fig. 2 and 3).

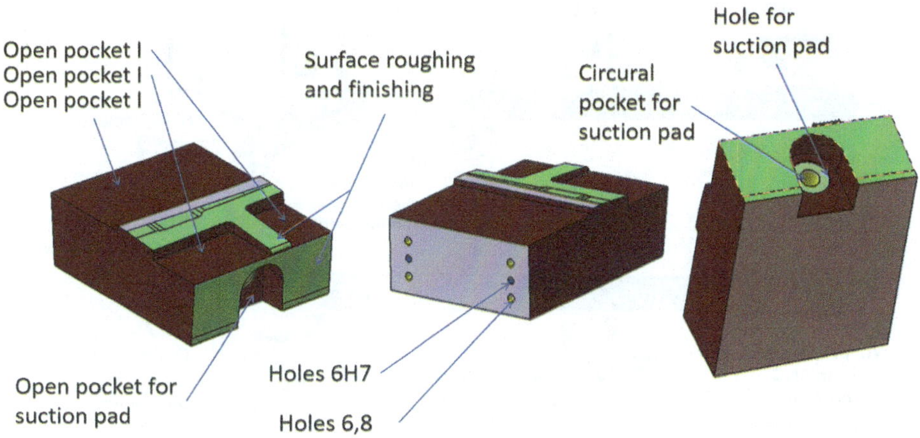

Fig. 3. Recognized geometric features of single welding tool

In the described case, this type of tooling is ordered by the customer who is responsible for its design. However, it has no influence on the method of its production in detail, as this is the subject of the manufacturer. The die group consists of 10 similar parts, so the manufacturer is interested in as efficient a process as possible (short time, less resource). One of the ways is to optimize it is to modernize the traditional approach in the CAM programming process, and for such need, the ACPUT procedure was developed.

4 Experiment and Results

To check the usability of the ACPUT procedure a comparing test to the traditional CAM programming method was proposed. For given group of welding tooling CAM programs were developed:

a) with the ACPUT procedure;
b) in the standard procedure by

   - a beginner programmer/technician,
   - an expert programmer/technician.

It should be emphasized that ACPUT (Fig. 4) is the procedure of implementation of CAM programming automation, not a specific IT tool, so the essence of the tests was to compare it with the traditional approach to programming (realized by technicians/CAM programmers), who can use any tools and methods in their work, even FR (Feature Recognition) or well-known GT assumptions (Group Technology).

The knowledge base, developed for the purposes of the ACPUT was also made available to the programmers (beginner and expert), so that the data for the machining process were the same in all cases.

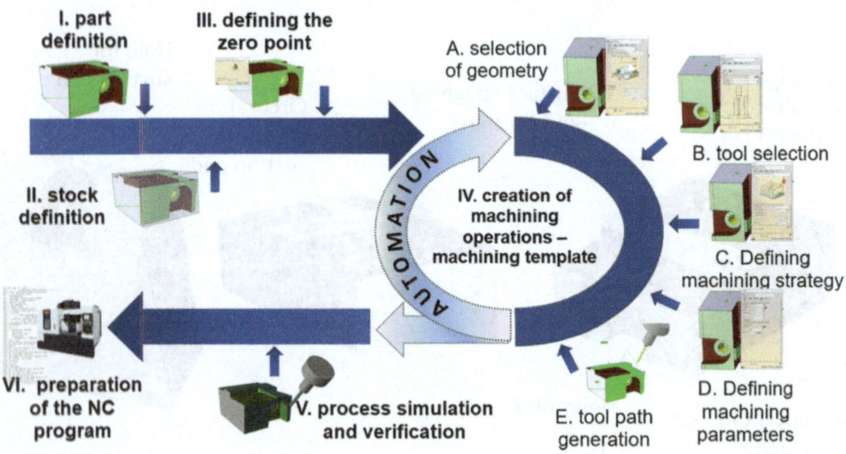

Fig. 4. Process of CAM programming for welding group of tooling using ACPUT procedure

The preparation times (cumulative) of the machining programs for all three cases are presented in Table 1. The automatic solution considered the time needed to develop the template (step 4 in Fig. 1 - it was added to the time for the first part), therefore the preparation time for the first part was much longer than the preparation times for the subsequent parts. A similar difference can also be seen in the work of the programmers programming in the traditional way. The differences between the expert and the beginner programmers' working times were related to their proficiency in using the CAM program.

The total preparation time (Tpt) for the machining programs for all the ten parts turned out to be the shortest in the ACPUT procedure and amounted to 193 min. This result was used as a reference to evaluate the other results. The working time of the beginner programmer was 478 min (representing 247% of the working time of the automatic solution), while the expert programmer had a working time of 298 min (154%of the working time of the automatic solution). The results presented in Table 1 are also shown in Fig. 4. This shows that the ACPUT procedure is already justified when the group of parts is greater than two (Fig. 5).

After execution and verification of the prepared machining programs, a simulation of the machining in the CAM system was performed, which made it possible to determine the machining times for the individual parts.

Table 1. Cumulative preparation time for machining programs

CAM preparation time [min] - cumulative			
After part no	ACPUT procedure	Standard procedure by beginner	Standard procedure by expert
1	66	143	48
2	82	208	80
3	97	256	105
4	114	288	135
5	122	315	157
6	140	353	191
7	147	379	211
8	166	420	243
9	180	447	270
10	**193**	**478**	**298**

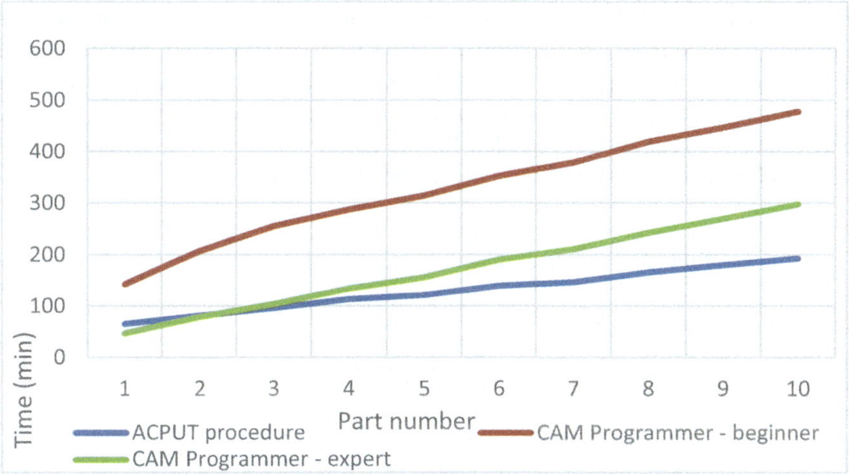

Fig. 5. CAM preparation time [min] - cumulative

Firstly, the obtained machining quality was checked. The measurements showed that in each of the three cases, the geometric accuracy, roughness, and surface and edge appearance on all the machined parts met the specified requirements. The cumulative summary of the time results from the simulation for each part separately is shown in Table 2.

Due to the fact that the parameters and the method of processing were specified in the technological knowledge base, and each of the programmer operated using the same data, the differences in the machining times for individual parts were insignificant. The difference between the ACPUT procedure and the expert programmer was only 0.6%,

Table 2. Cumulative machining time on CNC machine based on CAM simulation

Machining time [min] - cumulative			
After part no	ACPUT procedure	Standard procedure by beginner	Standard procedure by expert
1	131	146	125
2	327	326	335
3	492	504	493
4	713	730	724
5	959	984	964
6	1099	1146	1117
7	1307	1370	1313
8	1571	1625	1583
9	1763	1822	1769
10	**1883**	**1956**	**1895**

while between the ACPUT procedure and the beginner programmer it amounted to 3.9%. This mainly resulted from the selection of other speeds and the tool's path of movement during its approach and departure from the workpiece. It should be emphasized here that these parameters were not specified in the technological knowledge base.

The total time (cumulative) needed for the preparation of the machining program and the machining of individual parts is presented in Table 3.

Table 3. Cumulative time for manufacturing of tested parts

Machining time [min] – cumulative			
After part no	ACPUT procedure	Standard procedure by beginner	Standard procedure by expert
1	197	289	173
2	409	534	415
3	589	760	598
4	827	1018	859
5	1081	1299	1121
6	1239	1499	1308
7	1454	1749	1524
8	1737	2045	1826
9	1943	2269	2039
10	**2076**	**2434**	**2193**

The total time for the preparation of the machining programs and the machining for all ten pieces was 2076 min in the case of the programming using ACPUT procedure, 2434 min (representing 117.2% of the working time of the ACPUT procedure) for the beginner programmer and 2193 min (105.6% of the working time of the automatic solution) for the expert programmer.

5 Summary

Testing under industrial conditions showed that the ACPUT procedure makes it possible to reduce the time needed to develop a machining program. This has a positive effect on the total cost of tooling production.

The effectiveness of ACPUT was tested on the basis of the analysis of the time needed to prepare the CNC program and the machining operation time itself (based on the simulation in CAM). The preparation time of the program is a direct (next to the quality conditions) indicator of the effectiveness of ACPUT. Analyzing Table 1 and Fig. 4, it can be seen that the time benefits of using an automatic solution are significant, especially when compared to a less experienced programmer.

It should be emphasized that the effectiveness of the ACPUT procedure depends on the experience of those who prepare the templates. For the tests carried out two technicians were invited to participate: one with several years' experience and one with limited experience (working in this position for only a few months). The first was considered an expert and the second a beginner. The aim of the study was not to indicate the predictable differences between them, but to evaluate ACPUT which constituted a reference. Separating the programmers into more experienced and less experienced had one more purpose - to assess whether, after preparing the machining templates according to ACPUT, someone with less experience would be able to use the automatic solution successfully. The study revealed that it was possible: the results obtained using the automatic solution were even better than the work produced by expert.

References

1. Brettel, M., Friederichsen, N., Keller, M., Rosenberg, M.: How virtualization, decentralization and network building change the manufacturing landscape: an Industry 4.0 perspective. Int. J. Mech. Aerosp. Ind. Mechatron. Manuf. Eng. **8**, 1 (2014)
2. Hamrol, A., Gawlik, J., Sładek, J.: Mechanical engineering in Industry 4.0. Manag. Prod. Eng. Rev. **10**, 14–28 (2019)
3. Liu, Y., Xu, X.: Industry 4.0 and cloud manufacturing: a comparative analysis. J. Manuf. Sci. Eng. **139**(3) (2017)
4. Fogliatto, F.S., Da Silveira, G.J., Borenstein, D.: The mass customization decade: an updated review of the literature. Int. J. Prod. Econ. **138**(1), 14–25 (2012)
5. Zawadzki, P., Żywicki, K.: Smart product design and production control for effective mass customization in the Industry 4.0 concept. Manag. Prod. Eng. Rev. **7**(3), 105–112 (2016)
6. Zawadzki, P.: Methodology of KBE system development for automated design of multivariant products. In: Hamrol, A., Ciszak, O., Legutko, S., Jurczyk, M. (eds.) Advances in Manufacturing. LNME, pp. 239–248. Springer, Cham (2018). https://doi.org/10.1007/978-3-319-68619-6_23

7. Kowalski, M., Zawadzki, P.: Tooling CAD models preparation process for automated technology design system. In: Hamrol, A., Kujawińska, A., Barraza, M.F.S. (eds.) MANUFACTURING 2019. LNME, pp. 36–44. Springer, Cham (2019). https://doi.org/10.1007/978-3-030-18789-7_4
8. Kowalski, M., Zawadzki, P.: Decomposition of knowledge for automatic programming of CNC machines. Manag. Prod. Eng. Rev. **10** (2019)
9. Kowalski, M.: Method of automatic CAM programming using machining templates. Mechanik **93**(1), 48–52 (2020). https://doi.org/10.17814/mechanik.2020.1.4
10. Kowalski, M., Zawadzki, P., Hamrol, A.: Effectiveness of automatic CAM programming using machining templates for the manufacture of special production tooling. Strojniški Vestnik – J. Mech. Eng. **67**(10), 475–488 (2021). https://doi.org/10.5545/sv-jme.2021.7285
11. Babic, B., Nesic, N., Miljkovic, Z.: A review of automated feature recognition with rule-based pattern recognition. Comput. Ind. **59**(4), 321–337 (2008)
12. Zhang, Z., Jaiswal, P., Rai, R.: FeatureNet: machining feature recognition based on 3D convolution neural network. Comput. Aided Des. **101**, 12–22 (2018)
13. Sateesh, P., Mahesh, P.V.: A methodology for feature extraction and recognition for CAD/CAM integration using step file. Int. J. Res. Innov. (IJRI) **4**(1), 711–725 (2017)
14. Skvortsov, V., Proletarsky, A., Arzybaev, A.: Feature recognition module of the CAPP system. In: 2019 IEEE Conference of Russian Young Researchers in Electrical and Electronic Engineering (EIConRus), pp. 1769–1772. IEEE, January 2019
15. Zhou, G., Yang, X., Zhang, C., Li, Z., Xiao, Z.: Deep learning enabled cutting tool selection for special-shaped machining features of complex products. Adv. Eng. Softw. **133**, 1–11 (2019)
16. Tan, C., Ismail, N.: Design of a feature recognition system for CAD/CAM integration. World Appl. Sci. J. **21**(8), 1162–1166 (2013)
17. Sivakumar, S., Dhanalakshmi, V.: An approach towards the integration of CAD/CAM/CAI through STEP file using feature extraction for cylindrical parts. Int. J. Comput. Integr. Manuf. **26**(6), 561–570 (2013)
18. Miao, H.K., Sridharan, N., Shah, J.J.: CAD-CAM integration using machining features. Int. J. Comput. Integr. Manuf. **15**(4), 296–318 (2002)

On the Machining of Aluminum Alloy Series 7 with EDM

Maria Balanou[1], Panagiotis Karmiris-Obratański[1,2], Nikolaos E. Karkalos[1], Emmanouil L. Papazoglou[1], and Angelos P. Markopoulos[1](✉)

[1] School of Mechanical Engineering, Laboratory of Manufacturing Technology, National Technical University of Athens, Athens, Greece
amark@mail.ntua.gr

[2] Faculty of Mechanical Engineering and Robotics, Department of Manufacturing Systems, AGH University of Science and Technology, Cracow, Poland

Abstract. Electrical Discharge Machining is a thermo-electrical non-conventional machining process, which is widely used in machining difficult-to-cut materials in complex shapes and geometries, regardless their mechanical properties. Aluminum series 7 alloys are the strongest wrought aluminum alloy with extensive utilization in numerous modern applications. In the current study, an experimental investigation of machining Al7075 with EDM and brass electrode is presented. Full-scale experiments were conducted, with control parameters the pulse-on current and the pulse-on time, while, as the machining performance indexes, the Material Removal Rate (MRR), and the Tool Material Removal Rate (TMRR) were chosen and Surface Roughness (SR) was estimated in terms of the maximum roughness Rz and the average surface roughness Ra. Finally, a statistical analysis was followed in order to assess how each machining parameter affects the machining results.

Keywords: Electrical discharge machining · Aluminum alloy · ANOVA

1 Introduction

Electrical Discharge Machining (EDM) has been established as one of the most widely utilized non-conventional manufacturing processes in machining electrically conductive materials. EDM consists of a feasible method for machining parts in high precision regardless of the geometrical complexity, having applicability from the macro to the micro-scale [1]. Some main engineering applications that EDM is applied are the manufacturing of surgical components, dies and mold-making parts, microelectronics, automotive, and aerospace industries [2, 3].

EDM is a thermo-electric machining process where the material removal is related to the erosive effect caused when repeated electric discharges are generated between two conductive electrodes [4]. During the EDM process, a voltage is applied between the working electrode and the workpiece that both are immersed in a dielectric medium. At the closest points between the two electrodes, the dielectric fluid is ionized, and a

B. Gapiński et al. (Eds.): MANUFACTURING 2022, LNME, pp. 149–160, 2022.
https://doi.org/10.1007/978-3-031-00805-4_13

spark discharge is generated. From each electric spark, high thermal energy densities are produced, sufficient to melt and/or vaporize material from both the electrode and the workpiece. A portion of the aforementioned melted material is removed, resulting in the formation of the workpiece to the desired shape, while at the same time, in the wear of the tool electrode as well.

EDM is a complex machining process with several process parameters such as the pulse-on current (I_P), pulse-on time (T_{on}), dielectric fluid etc., that affect the performance indexes, such as the Material Removal Rate (MRR), the Tool Material Removal Rate (TMRR), the Surface Roughness (SR), and Surface Integrity (SI). Chandramouli and Eswaraiah [5] conducted an experimental investigation in order to determine the optimal process parameters of MRR and SR during EDM by employing the Taguchi Design of Experiment (DOE) method and an Analysis of Variance (ANOVA). The workpiece material was 17-4 PH Steel, while a copper-tungsten electrode was utilized. The process parameters were the I_P, the T_{on}, the pulse-off time (T_{off}), and the tool lift time. The results showed that the I_P and the T_{on} had the dominant influence on MRR and SR. Singh et al. [6] machined a stainless steel 316L using a copper electrode. Pulse-on current, pulse-on time, pulse-off time, and voltage were selected as machining parameters, and the Taguchi method was again used. The relationship between machining parameters and output measures (MRR and TMRR) was analyzed using Regression Model Analysis. They revealed that pulse-on current affects MRR and TMRR, with an increase in I_P leads to increased MRR and TMRR. The T_{on} was the most significant factor which influenced the SR of the machined workpiece. Sahoo et al. [7] analyzed MRR and TWR on Inconel 625 using a copper electrode. The Taguchi method and ANOVA were used, and the process control parameters were the T_{on}, the flushing pressure, and the I_P. Moreover, the utility method was used for the optimization of the response variables, aiming to maximize the MRR and minimize the TWR during machining. Mohanty et al. [8] investigated the possibility of surface modification by using mixed powder dielectric and the ability of surface hydrophobicity by using the μ-EDM process. The results show that surface hydrophobicity is inducted as the water contact angle with the surface is between 91.28° to 118.23°.

By analyzing the available literature, it was deduced that most EDM process works focused on different machining types of steel alloys, and very limited research has been reported on the machining performance of aluminum and its alloys. Aluminum (Al) alloys are widely used in many industries such as automotive, marine, and aerospace because of their low weight, corrosion resistance, and high strength. Although aluminum alloys have high machinability with the conventional machining process, EDM is usually used for the production of complex shapes of Al components with extremely high accuracy. Moreover, considering also that no mechanical forces are exerted on the workpiece during EDM, this process can be definitely regarded as an alternative for machining Al alloys. Thus, the EDM process is used for machining products, such as mold components, dies, structural components for the aerospace industry, and forming tools [9].

Arooj et al. [10] performed experiments for Al 6061 T6 workpiece using a copper electrode. Pulse-on current was chosen as a machining parameter for the conducted experiments. According to the deduced conclusions, MRR, SR, and the average white layer thickness were increased as the I_P had higher values. Ming et al. [11] conducted experiments to study the distribution of energy in the workpiece during the cutting of the Al 6061, Inconel 718, and SKD 11 under different EDM parameters, including the I_P and the T_{on}. The experimental results indicated that the proportion of the transferred discharge energy to the workpiece is very small. This fraction of the energy is associated with the pulse-on current and pulse duration and it varies from 7.998 to 12.005%, 2.879 up to 3.485%, and 2.976 up to 3.716% for Al6061, Inconel718, and SKD11, respectively. Moreover, the optimization of the process control parameters was investigated, considering the MRR and the discharge energy efficiency simultaneously. Niamat et al. [12] investigated the effect of T_{on} and I_P on MRR, TMRR, and microstructure analysis using distilled water and kerosene as dielectrics. The workpiece material was Al 6061 T6 alloy, and the electrode was graphite. The results showed that higher TMRR and MRR were achieved at higher values of I_P and T_{on} with distilled water. Microstructure analysis revealed that higher values of current produced surfaces with increased roughness. Routara et al. [13] analyzed MRR, TMRR, and SR on T6-Al 7075 alloy using Cu electrodes in steady and rotary condition. The spark gap, the Toff, and the IP were selected as control parameters for the carried-out experiments. TOPSIS method was used for the optimization of process parameters. It was deduced that the T_{off} and spark gap were the significant parameters for MRR in both, the rotary and steady mode of the tool. Moreover, the pulse-off time and pulse-on current affected more the surface roughness parameters for both the machining condition. Bédard et al. [14] analyzed the MRR, EWR, and Sa on machining A6061 alloy using a copper electrode and it was concluded that the MRR was affected by the discharge current and the discharge duration, while EWR was affected by these parameters, and by the delay between two series of discharges as well. Further, the Sa was influenced by the discharge current, the discharge duration, the voltage, the delay between two discharges, and the delay between two series of discharges. In addition, a number of studies have been carried out regarding the machining with EDM of composite materials that have Al alloy as the base matrix. Some examples of AMCs machined by EDM are Al-SiC [15], Al-B4C [16], etc. Finally, Tharian et al. [17] performed a multi-objective parametric optimization of machining an AA 7075 using WEDM. The results show that the most significant parameters affecting the MRR and the SR are the wire feed rate and the pulse-on current.

The present work experimentally investigates the machining of aluminum7075 alloy with EDM. The aluminum 7075 series (also called aluminum–zinc alloy) is the strongest wrought aluminum alloy and is widely used as a structural material in marine, automotive, and aviation applications due to its mechanical properties. The effect of pulse-on current and pulse-on time, on the MRR and TMRR has been investigated in order the efficiency of the machining to be evaluated. Moreover, the Surface Roughness was estimated in terms of the maximum roughness Rz and the average surface roughness Ra. Finally, a statistical analysis was followed, in order the impact of each control parameter on the machining results to be defined. The major contribution of the current study in the field of EDM pertains to the presentation of an experimental study regarding the machining of aluminum series 7 alloy with EDM, which, as far of authors best of knowledge, has been only briefly presented and discussed in the relevant literature.

2 Materials and Methods

The current series of experiments were carried out on a die-sinking EDM machine (AgieCharmilles Roboform 350 Sp). Aluminum alloy Al7075 was selected as a workpiece material because of its properties (high hardness, machinability, fatigue strength) and due to its wide application in aerospace, automotive industries, etc. The chemical composition of Al7075 is listed in Table 1. A brass electrode was utilized as a working tool and kerosene as a dielectric medium.

In the current experimental study, two machining control parameters, i.e., the pulse-on current and the pulse-on-time, were considered as input parameters, each of one having four levels. A full-scale series of experiments were conducted, while the Material Removal Rate (MRR), the Tool Wear Rate (TWR), and the Surface Roughness (SR) were chosen as the most significant machining performance indexes, and thus they were measured and statistically post-analyzed. In Table 2 the specific machining parameters are listed, along with the constant parameters (e.g., the duty factor, the open-circuit voltage, etc.) The pulse-on current varied from 5 to 17 A, and the pulse-on time from 12.8 to 100 μs covering a sufficient range of machining power and per pulse energies. Finally, the nominal cut depth was set at 1 mm to ensure a fully developed and steady-state machining process.

Table 1. Chemical composition, of Al-7075 (wt.%).

Al	Zn	Si (max)	Fe (max)	Ti (max)
Bal	5.1–6.1	0.4	0.5	0.2
Cu	Mn (max)	Mg	Cr	Zr + Ti (max)
1.2–2	0.3	2.1–2.9	0.18–0.28	0.25

Table 2. Input parameters.

Parameters	Level 1	Level 2	Level 3	Level 4
Pulse-on current (I_p)	5 [A]	9 [A]	13 [A]	17 [A]
Pulse-on duration (T_{on})	12.8 [μs]	25 [μs]	50 [μs]	100 [μs]
Duty factor	0.5			
Polarity	Straight			
Open circuit voltage	120 [V]			
Dielectric Fluid	Cactus normal paraffin and hydrocarbon			

The MRR and TMRR were defined by measuring the weight difference of the workpiece and tool material before and after EDM for a given machining time, using Eq. (1) and Eq. (2) respectively.

$$MRR = \frac{W_{wb} - W_{wa}}{t \cdot \rho_w} \quad (1)$$

$$TMRR = \frac{W_{eb} - W_{ea}}{t \cdot \rho_e} \quad (2)$$

where W_{wb} and W_{wa} are the weight of the workpiece before and after the machining in gr, W_{eb} and W_{ea} are the weight of the electrode before and after machining in gr, t the machining time in min, while ρ_w and ρ_e the workpiece and electrode density respectively. The workpiece and electrode densities are 0.00281 and 0.00873 gr/mm^3 respectively.

The surface roughness was evaluated by utilizing a VHX-7000 ultra-deep-field microscope (KEYENCE, Mechelen, Belgium) according to Focus Variation Microscopy (FVM) and a 20–2000 objective. This technology is comparable to confocal microscopy, and the measuring mechanism of an FV microscope is based on a white light LED source that goes via a semi-transparent mirror and a lens before reaching the measuring surface. The reflected light from the targeted regions then returns through the lens and is directed by a beam splitter onto a photonic detector, which records the geometric and photometric information. In other words, by combining the shallow depth of a traditional optical system with vertical scanning, FV may offer colorful high-resolution 3D surface data. After conducting the experiments, the SR was determined from 2D photos based on the ISO 25178 norm.

3 Results

The experimental results are presented in Table 3.

3.1 Material Removal Rate and Tool Material Removal Rate

It is well known and widely accepted that the EDM process results in highly nonlinear trends, and hence, the impact of the machining parameters (i.e., IP and Ton) must be carefully and thoroughly studied. The Ip and the Ton are related to the machining power and the per-pulse discharged energy. Hence, instinctively, an assumption can be made that the higher pulse-on current and/or pulse-on time leads to higher MRR and/or higher TMRR. Nevertheless, this hypothesis is not necessarily correct, and in fact, in many times, it is entirely wrong, leading in misconceptions. Although the higher nominal machining power and discharged per-pulse energy may result in increased molten material volume, complex underlying physical mechanisms strongly affect the overall process, namely, the plasma expansion over time, the debris presence and density and the carbon deposition. In more details, as the pulse-on time increases, the plasma channel expands consuming significant proportion of the discharged energy, while, the power is distributed in a larger area, leading in lower power densities. In parallel, higher machining power and/or per-pulse energy due to increased Ip and/or Ton may increase

Table 3. Experimental results

I_p [A]	T_{on} [μs]	MRR [mm^3/min]	TMRR [mm^3/min]	Ra [μm]	Rz [μm]
5	12.8	3.71	0.81	1.39	11.00
5	25	2.81	0.66	2.03	12.69
5	50	1.89	0.48	2.39	14.47
5	100	1.18	0.33	2.39	14.48
9	12.8	7.94	1.52	2.34	13.64
9	25	7.51	1.38	2.38	14.22
9	50	5.02	0.97	2.56	16.96
9	100	3.28	0.69	4.48	21.24
13	12.8	25.42	4.24	3.85	22.34
13	25	31.72	4.64	4.66	24.20
13	50	20.04	3.29	5.05	27.41
13	100	16.40	2.45	7.74	36.09
17	12.8	29.23	4.77	5.06	25.31
17	25	31.22	4.66	5.85	31.82
17	50	29.99	4.38	7.24	34.24
17	100	25.64	3.59	8.20	39.59

the debris concentration in between the workpiece and the electrode. The presence of these debris significantly influences the process since they do not only interact with the plasma channel and the occurred spark (in extreme cases even arcing conditions may be arisen), but, at the same time, they consume energy for their re-melting. Another phenomenon that takes place and affects the efficiency of the process is the carbon deposition on the electrode and the workpiece. Carbon that is coming from the decomposition of the hydrocarbon oil due to the extremely high topical temperatures, is deposited on the surfaces forming a carbon-based layer. At the same time, this layer may act as a protective layer for the electrode and negatively for the MRR. From this brief analysis, it is reasonably concluded and deduced that the MRR and the TMRR are expected not to change linearly and monotonously regarding the pulse-on current and the pulse-on time.

In Fig. 1 the Mean Effects Plot and the Interaction Plot of the MRR are presented. From the Main Effects Plots analysis, it is easily deduced that the IP is the machining parameter that mainly affects the MRR, whilst the pulse-on time has a minor and vague impact. More specifically, the mean MRR increased by 294% between 9 and 13 A, while the mean MRR, at first increased (from 12.8 to 25 μs), but further increase of Ton resulted in its decrease. Nevertheless, and in order to avoid any false deduced conclusions, it is absolutely necessary to study the Interaction Plot closely. The first interesting observation pertains to the deviation in the MRR values for the same Ips. The deviation for 5, 13, and 17 A is very low, while, for 13 A becomes significant. Another remark is that a

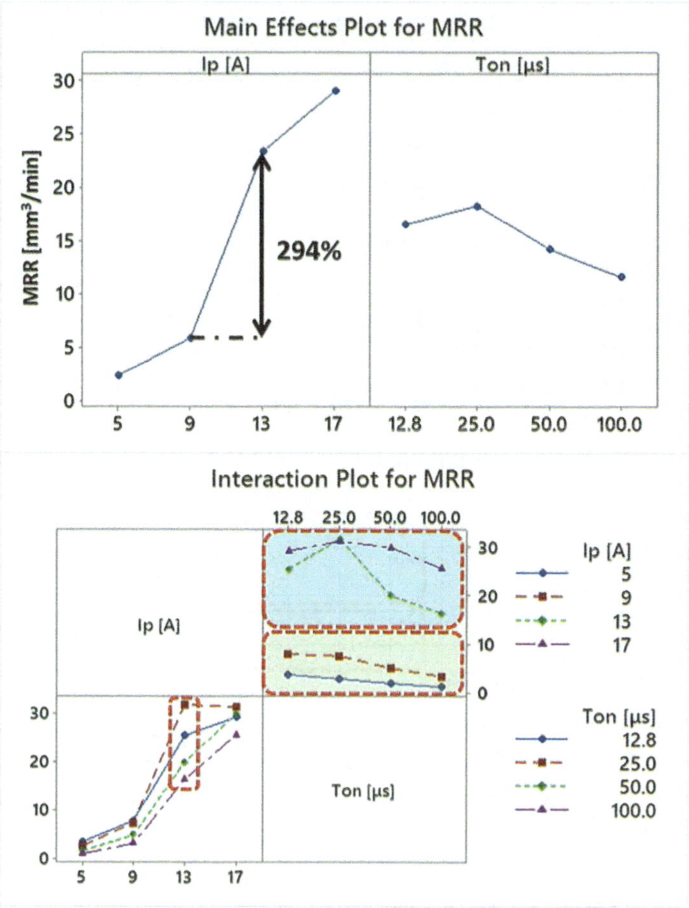

Fig. 1. Main effects plot and interaction plot for MRR.

13 A increase in Ton from 25 to 100 μs resulted in a significant decrease of the MRR. In general, the higher Ton seems to negatively affect the MRR, particularly for pulse-on current 13 A. A final point that is worth to be mentioned is the clear differentiation between 9 and 13 A. Although the pulse-on current values have equidistant from each other, the shift from 9 to 13 A resulted in a notable increase of MRR, comparatively with the change between 5 and 9 A or 13 and 17 A. As an overall conclusion and assessment, it can be said that the experimental results of MRR are in accordance with the theory regarding the effect of pulse-on current and pulse-on time on the process, whereas, the process outcome is strongly affected by the combination of the machining parameters and not each one separately.

In Fig. 2 the Mean Effects Plot and the Interaction Plot of the TMRR are presented. By studying the Main Effects Plot of TMRR we find that MRR and TMRR follow the same trend. Namely, the Ip is the parameter that mainly affects the mean value of TMRR, while, the pulse-on time seems to have a less important effect. Between 9 and 13 A the

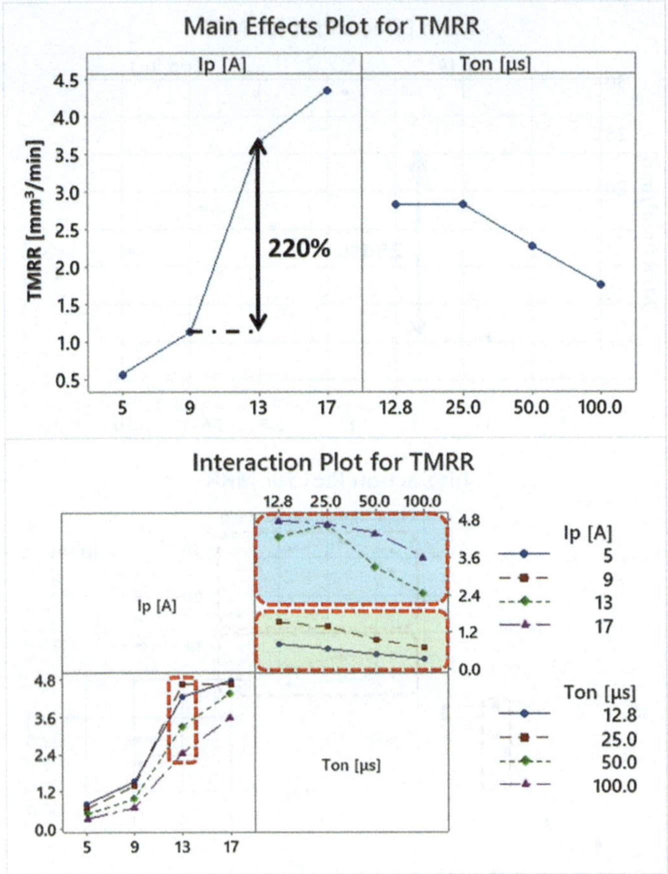

Fig. 2. Main effects plot and interaction plot for TMRR.

mean TMRR increased by 220%, whilst, for Tons higher than 25 μs the mean TMRR shows a decrease trend. Nevertheless, a more in-depth analysis is necessary, hence, the Interaction plot has to be carefully considered. From the Interaction Plot of TMRR it is deduced that higher Ip leads in higher TMRR, while, in most cases, higher Ton results in lower TMRR. It is also noticeable that again for 13 A there is a significant deviation in TMRR values depending on the Ton and the differentiation between low and high-power machining (green and blue areas, respectively). The importance of the machining parameters combination regarding the TMRR is shown from the above observations. Considering that the efficiency of the process depends on the ratio of the TMRR to the MRR, it has to be pointed out that higher TMRR does not compulsorily result in less efficient machining. For example, for the lower TMRR (5 A - 100 μs - 0.38 mm/min), the MRR to TMRR is 0.28 while for the higher TMRR (17 A - 12.8 μs - 5.44 mm/min) the respective ratio is 0.16 implying far more efficient machining. Hence, it is easily deduced that the combination of the machining parameters has to be considered during

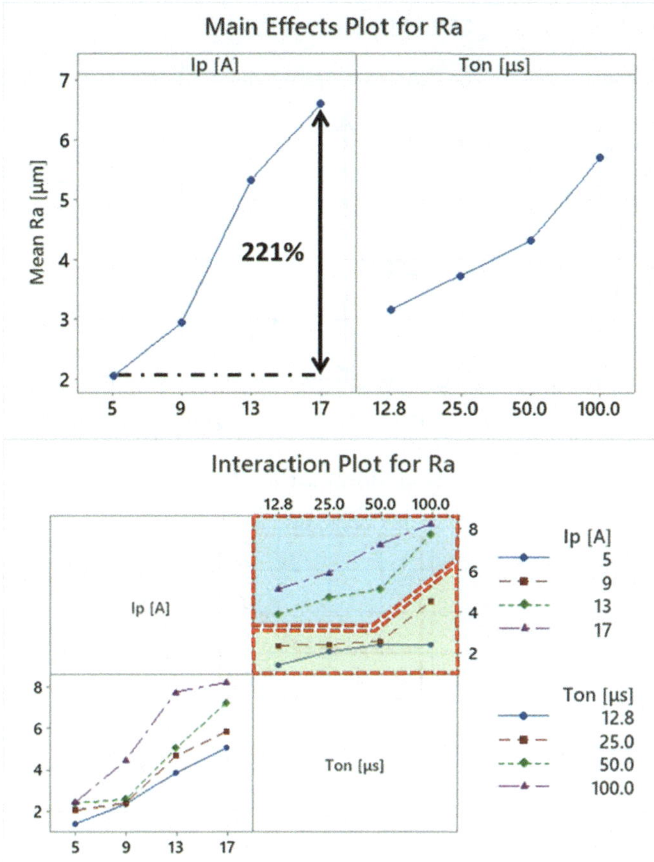

Fig. 3. Main effects plot and interaction plot for Ra.

the process planning, along with any other requirements, e.g., Surface Quality and/or Surface Roughness.

3.2 Surface Roughness

The morphological characteristics of the machined surfaces are the result of successive high-energy spark that acted on the surfaces. The surface roughness of the EDMed surface is correlated with the distribution of the craters formed due to the removal of material by repeated electric sparks. Since the EDM surface is mainly created due to the discharge energy, I_P and T_{on} are two of the most significant factors which affect the size of the produced crater, i.e. the SR. As a rule of thumb, by an increase in pulse-on time, the period of transferring the discharge energy into the electrodes and the plasma column are increased, leading to larger diameter craters. On the other hand, the high pulse-on current results in the generation of high discharge energy which creates impact forces and causes deeper craters on the machined surface [18]. The surface roughness is

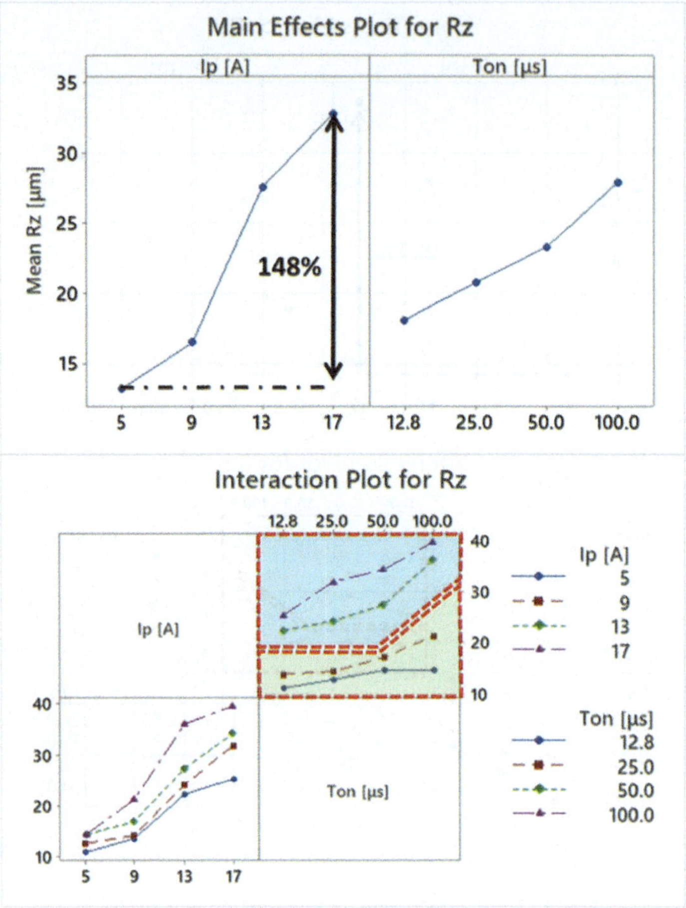

Fig. 4. Main effects plot and interaction plot for Rz.

estimated by the sum of the maximum valley depth and maximum peak height Rz, and an arithmetical mean of local peaks Ra.

In Fig. 3 and Fig. 4, the Main Effect plots and the Interaction plots for Ra and Rz are presented. The Main Effect plot of Fig. 3 implies that the Ra significantly increases as the I_P increases. On the other contrary, the change in T_{on} does not have the same clear effect on Ra and this is mainly due to the significant deviation for the same T_{on} in the Ra values. Namely, and based on the Interaction plot of Ra, for the same I_P, higher T_{on}s result in increased Ra. Nevertheless, for the same pulse-on times there is a very high deviation in Ra values depending on the I_P, implying and suggesting that the Ra is strongly affected by the combination of the machining parameters and the per-pulse energy. For example, the combinations of 13 A - 50 μs and 17 A - 12.8 μs have almost the same Ra, indicating that during the machining planning, and based on the SR requirements the combination of I_P and T_{on} has to be always considered in order the most efficient machining parameters combination to be selected. Another worth to be mentioned observation is that for the low energies machining the Ra seems to be practically constant. For example, for 9 A between 12.8 and 50 μs there is almost no change in Ra, as for 5 A between 25 and 100 μs. Finally, similar results are observed for Rz (see Fig. 4), and hence it would be verbalization a further extensive analysis.

4 Conclusion

In the current study an experimental investigation of machining aluminum alloy 7075 with EDM has been carried out. A full-scale series of experiments was conducted with two control parameters, the pulse-on current and the pulse-on time, whilst the rest of machining parameters were kept constant. The machining efficiency was assessed on the basis of the MRR and TMRR, whilst the SR in terms of Ra and Rz. The followed statistical analysis defined how each parameter affects the machining performances. Based on the obtained results and post-analysis, the following main conclusions were deduced:

- The MRR mainly is mainly affected by the I_P, while the T_{on} has a more fuzzy impact on it. More specifically, as the pulse-on current increased from 9 to 13 A the mean MRR increased by 294%.
- The TMRR is also mainly affected by the I_P, having an increase of 202% between 9 and 13 A, whilst again the T_{on} seems to have a vague impact.
- The process, in respect of the MRR and the TMRR can be classified in low-power machining and high-power machining, having a distinguished behavior.
- The pulse-on current has the major effect on the SR, as its increase results in higher Ra and Rz, while, the combination of machining parameters, and hence the per-pulse energy is also extremely significant.

References

1. Mohanty, S., Kumar Das, A., Dixit, A.R.: Influence of tool materials on surface modification using μEDC process. Surface Eng. **37**(9), 1084–1097 (2021). https://doi.org/10.1080/02670844.2021.1900518
2. Jahan, M.P. (ed.): Electrical Discharge Machining (EDM): Types, Technologies and Applications. Nova Science Publishers, New York (2015)
3. Ho, K.H., Newman, S.T.: State of the art electrical discharge machining (EDM). Int. J. Mach. Tools Manuf. **43**(13), 1287–1300 (2003). https://doi.org/10.1016/S0890-6955(03)00162-7
4. Liu, S., Thangaraj, M., Moiduddin, K., Al-Ahmari, A.M.: Influence of adaptive gap control mechanism and tool electrodes on machining titanium (Ti-6Al-4V) alloy in EDM process. Materials **15**, 513 (2022). https://doi.org/10.3390/ma15020513
5. Chandramouli, S., Eswaraiah, K.: Experimental investigation of EDM process parameters in machining of 17-4 PH Steel using taguchi method. Mater. Today Proc. **5**(2), 5058–5067 (2018). https://doi.org/10.1016/j.matpr.2017.12.084
6. Singh, G., Bhui, A.S., Lamichhane, Y., Mukhiya, P., Kumar, P., Thapa, B.: Machining performance and influence of process parameters on stainless steel 316L using die-sinker EDM with Cu tool. Mater. Today Proc. **18**, 2468–2476 (2019). https://doi.org/10.1016/j.matpr.2019.07.096

7. Kumar Sahoo, S., Thirupathi, N., Saraswathamma, K.: Experimental investigation and multi-objective optimization of die sink EDM process parameters on inconel-625 alloy by using utility function approach. Mater. Today Proc. **24**, 995–1005 (2020). https://doi.org/10.1016/j.matpr.2020.04.412
8. Mohanty, S., Kumar Das, A., Dixit, A.R.: Surface integrity of tribo-adaptive layer prepared on Ti6Al4V through μEDC process. Surface Coatings Technol. **429**, 127922 (2022). https://doi.org/10.1016/j.surfcoat.2021.127922. ISSN 0257-8972
9. Srinivasa Rao, P., Ramji, K., Satyanarayana, B.: Effect of wire EDM conditions on generation of residual stresses in machining of aluminum 2014 T6 alloy. Alexandria Eng. J. **55**(2), 1077–1084 (2016). https://doi.org/10.1016/j.aej.2016.03.014
10. Arooj, S., Shah, M., Sadiq, S., Jaffery, S.H.I., Khushnood, S.: Effect of current in the EDM machining of aluminum 6061 T6 and its effect on the surface morphology. Arab. J. Sci. Eng. **39**(5), 4187–4199 (2014). https://doi.org/10.1007/s13369-014-1020-z
11. Ming, W., et al.: Investigating the energy distribution of workpiece and optimizing process parameters during the EDM of Al6061, Inconel718, and SKD11. Int. J. Adv. Manuf. Technol. **92**(9–12), 4039–4056 (2017). https://doi.org/10.1007/s00170-017-0488-6
12. Niamat, M., Sarfraz, S., Shehab, E., Ismail, S.O., Khalid, Q.S.: Experimental characterization of electrical discharge machining of aluminum 6061 T6 alloy using different dielectrics. Arab. J. Sci. Eng. **44**(9), 8043–8052 (2019). https://doi.org/10.1007/s13369-019-03987-4
13. Routara, B.C., Das, D., Satpathy, M.P., Nanda, B.K., Sahoo, A.K., Singh, S.S.: Investigation on machining characteristics of T6-Al7075 during EDM with Cu tool in steady and rotary mode. Mater. Today Proc. **26**, 2143–2150 (2019). https://doi.org/10.1016/j.matpr.2020.02.462
14. Bédard, F., Jahazi, M., Songmene, V.: Die-sinking EDM of Al6061-T6: interactions between process parameters, process performance, and surface characteristics (2020)
15. Naik, S., Sabat, S., Das, S.R., Dhupal, D., Nanda, B.K.: Experimental investigation, parametric optimization, and cost analysis in EDM of aluminium-silicon carbide metal matrix composite. In: Deepak, B.B.V.L., Parhi, D.R.K., Biswal, B.B. (eds.) Advanced Manufacturing Systems and Innovative Product Design. LNME, pp. 175–187. Springer, Singapore (2021). https://doi.org/10.1007/978-981-15-9853-1_15
16. Gopalakannan, S., Senthilvelan, T., Ranganathan, S.: Modeling and optimization of EDM process parameters on machining of Al7075-B4C MMC using RSM. Procedia Eng. **38**, 685–690 (2012). https://doi.org/10.1016/j.proeng.2012.06.086
17. Tharian, B.K., Jacob, E., Johnson, J., Hari, V.: Multi-objective parametric optimization in EDM using grey relational analysis. Mater. Today: Proc. **16**, Part 2, 1013–1019 (2019). https://doi.org/10.1016/j.matpr.2019.05.190. ISSN 2214-7853
18. Shabgard, M., Seyedzavvar, M., Oliaei, S.N.B.: Influence of input parameters on the characteristics of the EDM process. Stroj. Vestnik/J. Mech. Eng. **57**(9), 689–696 (2011). https://doi.org/10.5545/sv-jme.2011.035

Surface Structurization of Metallic Inserts for Enhancement of Polymer-Metal Joints

Bartosz Korzeniewski, Natalia Łożykowska, and Karol Bula(✉)

Institute of Materials Technology, Poznan University of Technology, Poznan, Poland
karol.bula@put.poznan.pl

Abstract. In this work, aluminum inserts were structurized and joined with polyamide 6 with 15% by weight of short glass fibres (PA6 GF15) by using the injection-molding technique. The metal parts, made of aluminum grade 6061 were surface structurized by means of plastic forming with shaped inserts and also by laser microablation. Plastic forming was made at the preset pressing pressure values: 5 MPa, 10MPa, 15 MPa and 18 MPa (in two-side mode) meanwhile laser beam speed was set as: 75, 150, 250 and 350 mm/s with grooves interline distance of 0,2; 0,4 and 0,8 mm (for one side and both sides regime). The joints made of aluminum/PA6 GF15 were mechanically tested in a tensile-lap-shear test. For estimation of the basic shape and parameters of grooves contact surface roughness tester was used. Microscopic observation with optical microscope confirmed that during injection stage polymer materials filled the grooves shape in both plastic formed as well as laser treated aluminum inserts. The maximal values of the joint's failure force under tensile-shear tests were achieved for two-side laser micropatterned samples (for scanning speed 150 mm/s and interline distance 0,4 mm) and it was 1494 N. The ultimate force detected for samples with plastic formed inserts was almost four times smaller and it was 325 N.

Keywords: Polyamide 6 · Aluminum · Hybrid joint · Overmolding · Laser engraving · Surface structurization

1 Introduction

Combining metals with polimers in one part offers manufacturing of lightweight and easily shaped elements, that are yet attractive in engineering applications. These opportunity is gained by the possibility of product forming in one single step process – overmolding, that avoid addition of a third component – adhesive [1, 2]. The formation of a strength joint of these two incompatible materials, without chemical affinity is augmented by the ability of molten polymer to infiltrate the metal surface [3, 4]. The infiltration of metallic surface by the molten polymer that flows into the mold cavity is preferred for surfaces that were previously microstructurized. Conducted tests showed that prepared metallic surface and its topography also play important role in micromechanical adhesion [5]. A lot of applications could be found for the inserts that are structurised by plastic working and microstructurised inserts e.g. toughening profile in the car armrest and in

B. Gapiński et al. (Eds.): MANUFACTURING 2022, LNME, pp. 161–174, 2022.
https://doi.org/10.1007/978-3-031-00805-4_14

an electrotechnical connectors used in electric appliances, accordingly. The enhancement of surface area and surface geometries could be created by using sandblasting micro abrasion [6], chemical etching [7] and by the plasma-etching pretreatment [8]. Recently, a lot of papers were dedicated to laser-micro-patterning as novel method used to develop of surface of metal inserts, subjected to overmolding by injection technique [8–12]. Rodríguez-Vidal et al. have broadly analyzed the laser beam working parameters on the formation of micro-pattern geometries as well as grooves geometry and formation of recast material. They have found correlation between depth and width of the grooves that might promote a great final joint strength [13, 14]. Generally speaking, the success of joining polymer to metal in overmolding technique is ruled on one hand by the surface structurisation, on the other hand equally important is to maintain ability to infiltration of that surface by the molten polymer. The last condition is governed by the insert surface temperature, temperature of polymer crystallization from molten state, polymer viscosity at injection temperature, injection speed and pressure.

1.1 Research Problem

In this work we present a comparative studies of two different approaches in structurization of aluminum. One of chosen methods is based on plastic forming/working by using hard tool inserts with diamond-like edgings on their surfaces and hydraulic press while the second one uses pulsed fiber laser to engrave a parallel grooves on metallic parts. Our purpose was to validate which one of this two kind of fast, waste less forming methods will offer suitable surface parameters to gain the polymer-metal micromechanical adhesion, reveal as an ultimate force applied to disconnect the joint. The tensile data will also be considered in the relation to roughness parameters of the metallic surfaces after structurization with contact (plastic forming) and non-contact (laser treatment) methods.

2 Experimental

For the production of lap joints in the overmolding technology, polyamide 6 filled with 15% by weight of glass fiber was used (Tarnamid® T27 GF15). As a metallic part an aluminum alloy 6061, often used in the automotive industry for stamped products was applied. The aluminum alloy 6061 sheets were cut and plates with 75 mm length, 10 mm width and 1 mm of thickness were subjected to structurization stage. For structuring through plastic forming/working an aluminum alloy 6061 plate was placed between the inserts with diamond-like edgings (Fig. 1), and then pressed with the hydraulic press. Four series of compression samples were made for the preset pressing pressure values (oil pressure in the piston actuator system): 5 MPa, 10 MPa, 15 MPa and 18 MPa.

Separate sheets of aluminum alloy 6061 were surface structured by using a 30W fiber laser, operating in pulsed mode. The laser operating parameters used in the experiment are listed in Table 1. Laser beam etched series of parallel, straight lines, which were aligned perpendicularly (in relation to force applied in later shear testing). To determine the effect of laser beam scanning speed and interline distance as independent factors, various samples were prepared and listed in Table 2.

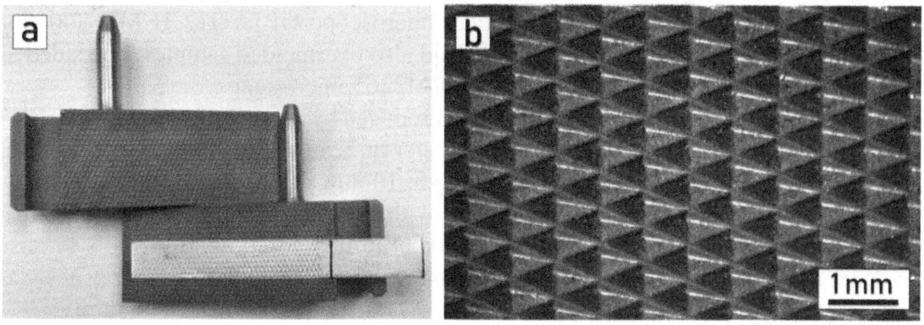

Fig. 1. Pictures of: press inserts: (a) press insert surface, (b) structurized surface of aluminum alloy 6061 plate

Table 1. Laser operating parameters.

Parameter	Unit	Value
Wavelength	nm	1064
Frequency	kHz	30
Beam intensity	W	30
Spot size	μm	<10
Scanning speed	mm/s	75; 150; 250; 350
Interline distance	mm	0,2; 0,4; 0,8

Table 2. Laser engraving – sample variants.

Scanning speed [mm/s]	Interline distance [mm]		
	0,2	0,4	0,8
75		X	
150		X	
250	X	X	X
350		X	

Engraved inserts had their surfaces scanned with surface roughness tester SRT-220 in order to measure the shape and parameters of grooves and recast material, and further correlate results with latter microscopic observations (for roughness test parameters, see Table 8 in Appendix). Finally, same measurements were made on actual joints, after destructive tests had been done, to check how dual-material coupling cooperated during shear load.

For microscopic observations an optical microscope OPTA TECH MB 200 was chosen. The cross sections of metallic parts and also overmolded samples (one sided as well as two sided) were prepared with Leica RM2265 microtome.

The polymer – metal hybrid samples were produced by means of ENGEL 20/80 HLS injection molding machine, with a screw diameter of 22 mm. The mold with cavities of the same flow length (150 mm total flow length, 10 mm width), and with its thickness of 2 mm and 3 mm was utilized. Metal plates with structurized surfaces were placed at the cavity, opposite to the gate position, and kept in the closed mold for 30 s to raise the temperature to 60 °C. Subsequently, filling of the cavity with PA6 composites was realized at 80 °C of mold temperature, with an average injection rate of 90 mm/s.

Tensile tests were completed according to EN ISO 1465 standard, by means of a Zwick Roel Z010 (Germany) universal testing machine, operating under AllSuite software, with a 10 kN load cell. The samples, with a single lap joint geometry were designated to determine the shear strength of joints by shear test, realized by tensile force, with a crosshead speed of 2 mm/min.

3 Results

3.1 Optical Microscopy Observation

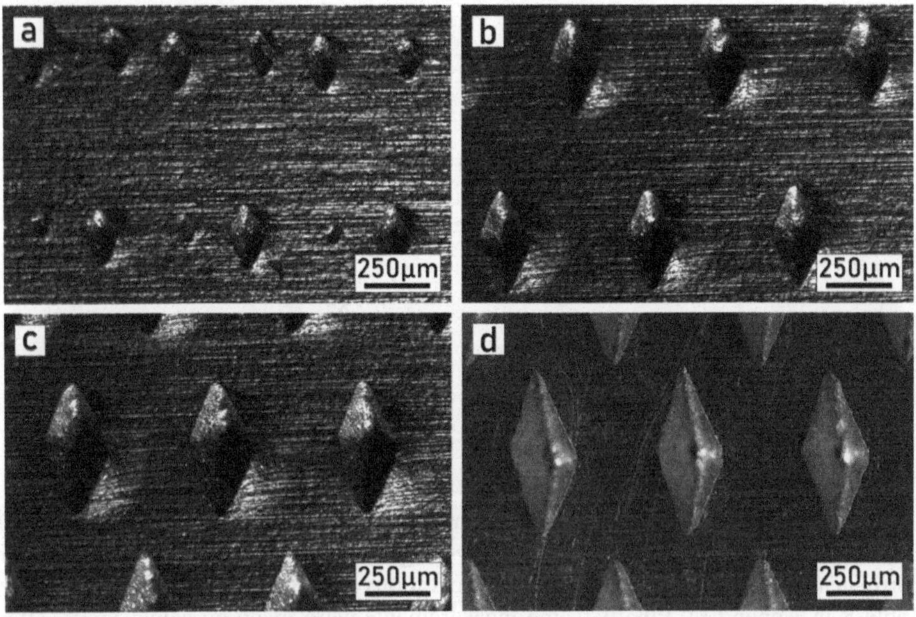

Fig. 2. Surfaces of plastic formed samples in relation to hydraulic press pressure: 5 MPa (a); 10 MPa (b); 15 MPa (c); 18 MPa (d).

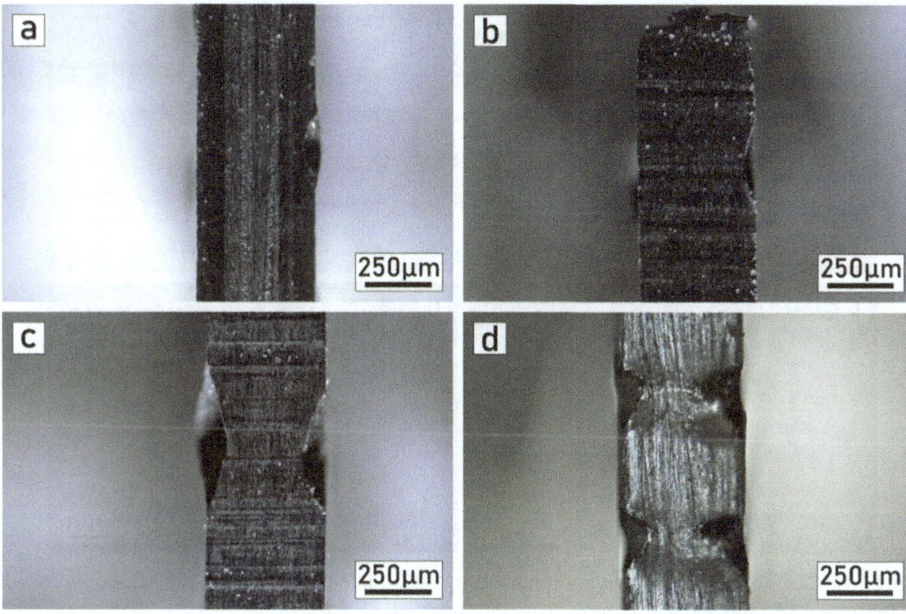

Fig. 3. Cross sectional view of plastic formed samples in relation to hydraulic press pressure: 5 MPa (a); 100 MPa (b); 15 MPa (c); 18 MPa (d).

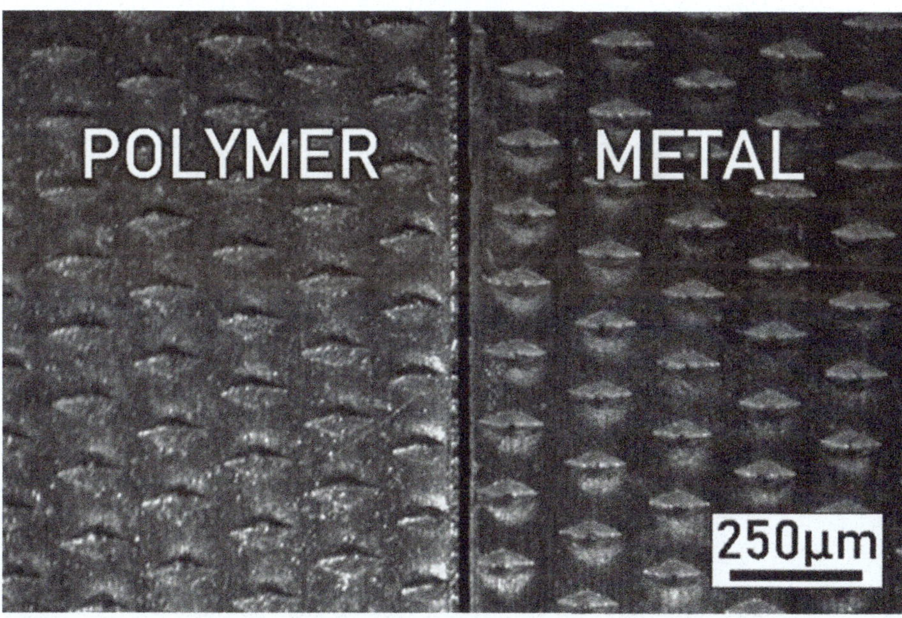

Fig. 4. Plastic formed aluminum alloy 6061 plate surface and corresponding polymer imprint.

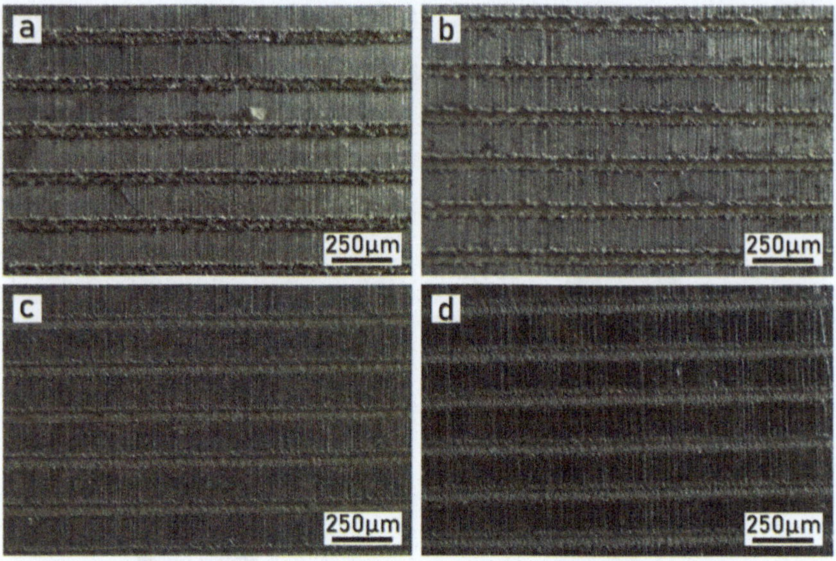

Fig. 5. Surfaces of laser engraved aluminum alloy 6061 plates in relation to laser scanning speed: 75 mm/s (a); 150 mm/s (b); 250 mm/s (c); 350 mm/s (d).

Fig. 6. Cross section of a one sided joint (laser engraving method).

Fig. 7. Cross section of a two sided joint (laser engraving method).

3.2 Roughness Parameters and Profiles of Structurized Al Plates

Table 3. Averaged roughness parameters of plastic formed Al alloy 6061 plates.

Symbol	Hydraulic press pressure [MPa]	Ra [μm]	Rz [μm]	Rsk [-]
a)	5	3,45	26,3	+0,034
b)	10	6,50	52,5	−1,148
c)	15	12,1	82,1	−2,041
d)	18	14,2	93,8	−1,700

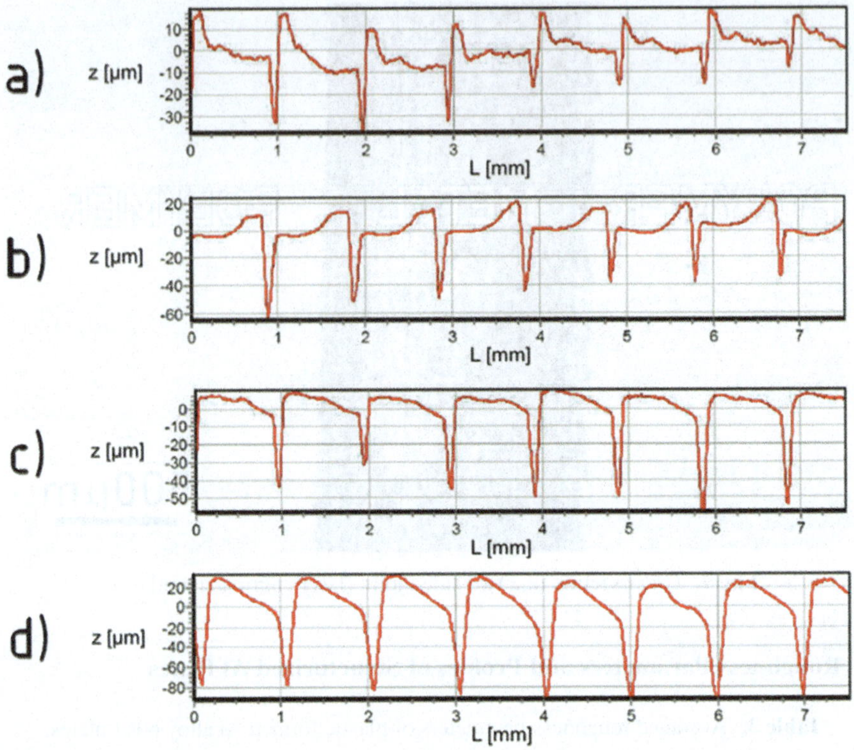

Fig. 8. Roughness profile curves of plastic formed samples at different pressures: 5 MPa (a); 100 MPa (b); 15 MPa (c); 18 MPa (d). Z – roughness profile curve height, L – sampling length.

Table 4. Average roughness parameters of laser engraved samples.

Scanning speed [mm/s]	Interline distance [mm]	Ra [μm]	Rz [μm]	Rsk [-]
75	0,4	16,3	54,5	+0,679
150	0,4	7,79	45,2	+1,034
250	0,2	38,7	6,68	−1,043
	0,4	3,48	36,1	−1,669
	0,8	2,72	43,8	−0,387
350	0,4	3,04	28,8	−1,891

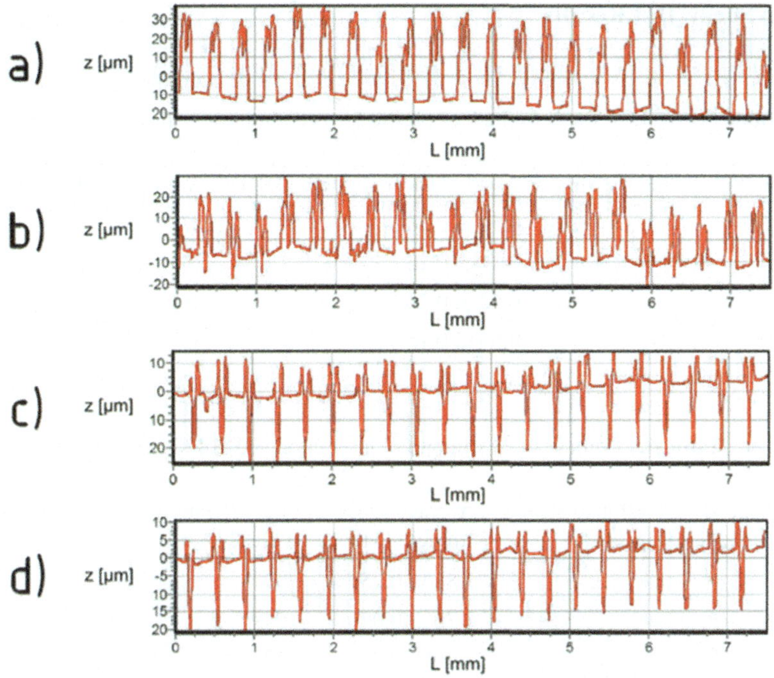

Fig. 9. Roughness profile curves of laser engraved samples at different scanning speeds: 75 mm/s (a); 150 mm/s (b); 250 mm/s (c); 350 mm/s (d). Z – roughness profile curve height, L – sampling length.

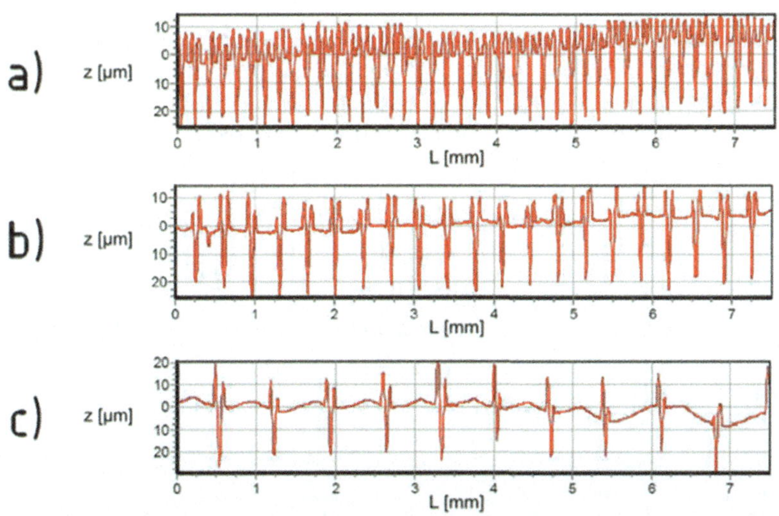

Fig. 10. Roughness profile curves of laser engraved samples with different interline distance: 0,2 mm (a); 0,4 mm (b); 0,8 mm (c). Z – roughness profile curve height, L – sampling length.

3.3 Lap Shear Test Results

Table 5. Average shear strength of plastic formed aluminum alloy 6061 plates.

Hydraulic press pressure [MPa]	Average shear strength [MPa]	Averagep failure force [N]
5	0,56 ± 0,28	274
10	0,31 ± 0,15	153
15	0,52 ± 0,26	254
18	0,66 ± 0,15	325

Table 6. Average shear strength of laser engraved one sided joints.

Scanning speed [mm/s]	Interline distance [mm]	Average shear strength [MPa]	Average failure force [N]	Number of series [-]
75	0,4	1,81 ± 0,25	973	7
150	0,4	1,76 ± 0,07	1049	7
250	0,2	1,40 ± 0,28	832	9
	0,4	1,15 ± 0,26	654	9
	0,8	0,87 ± 0,11	505	8
350	0,4	0,59 ± 0,16	360	8

Table 7. Average shear strength of laser engraved two sided joints.

Scanning speed [mm/s]	Interline distance [mm]	Average shear strength [MPa]	Average failure force [N]	Number of series [-]
75	0,4	1,16 ± 0,12	1230	8
150	0,4	1,41 ± 0,11	1494	8
250	0,2	1,36 ± 0,05	1443	6
	0,4	0,83 ± 0,39	874	7
	0,8	1,065 ± 0,110	1128	8
350	0,4	0,57 ± 0,31	630	3

4 Discussion

Figure 2 shows that surface indentations had pyramid shape and their size increased with the hydraulic press pressure. Also on the cross sections of aluminum alloy 6061 plates the triangle holes are visible (Fig. 3), especially for plates plastic formed under ultimate press pressure. This is further confirmed by surface roughness measurements (Table 3) and profiles (Fig. 8), where higher Ra, Rz and curve height indicate deeper and wider structures. Furthermore, as shown on roughness profiles, a kind of flash shape is formed due to plastic deformation of the surface.

Lap joints produced by injection overmolding had their surface indentations infiltrated by polymer material, as evident by the polymer imprint (Fig. 4), which was also confirmed for samples laser structured (Fig. 5).

Data from the shear test (Table 5) indicates that the higher the hydraulic press pressure, the better is shear strength of the joint, however this does not correlate with the lowest pressure series, which had the second highest strength. Failure force was analogous to above relationship. One of possible reason of that unusual situation is the specific, sharp character of roughness profile (Fig. 8a), that might suggest a better interlocking of polymer matrix around those indentations.

Lap joints of overmolded engraved plates show that polymer infiltrated exposed grooves and enveloped the recast material (Fig. 6 and 7). Shear test results (Table 6 and 7) show that increase in scanning speed does reduce the joint strength. However, the highest strength was achieved by intermediate speed, which included both high recast and partially exposed grooves. This indicates that both groove and recast have positive effect on joint strength.

Surface roughness profiles (Fig. 9 and 10) show that laser engraving parameters had significant effect on aluminum alloy 6061 plates' surfaces. Low beam scanning speed (75 and 150 mm/s) produced high recast on top of the grooves, obstructing them from the tester's probe and possibly restricted polymer infiltration in later process.

High beam scanning speed (250 and 350 mm/s) produces low recast and exposed grooves, but it is expected that the depth is smaller than real depth of low speed engravings, because higher speed correlates with lower heat output in the material.

Above relations are confirmed by Ra and Rz parameters decreasing with scanning speed (Table 4), where for low speeds (75 and 150 mm/s) decrease happened due to lowered recast height and for high speeds (250 and 350 mm/s) as a result of both recast height and groove depth reduction.

Discussed measurements also indicate that interline distance has low effect on groove and recast morphology, although the smaller the distance was, especially for 0,2 mm, the bigger change is. One should be underline here, that excessive energy output due to scanning lines being so close that heat left from previous scan does not dissipate in time and next line adds to the heat, probably leading to local thermal deformations.

Interline distance had visible effect on joint strength, but provided data is not enough to form relations yet, as intermediate distance yielded lowest strength.

One and two sided joints were compared. Data suggests that two sided joints offer more strength than one sided joints, but despite doubled area of joint surface, increase in strength is less than doubled.

Lastly, comparing shear test data of plastic formed and laser engraved samples shows that the laser method provides around three times the strength of plastic formed samples.

5 Conclusions

Both plastic forming and laser engraving surface structurization methods produced applicable structures and yielded positive results on joint strength in injection overmolding method. Microscopic observations confirmed that during injection stage polymer materials filled the grooves shape in plastic formed inserts as well as in laser treated aluminum alloy 6061 plates.

The ultimate force detected for samples with plastic formed inserts was almost four times smaller than in the case of the best laser structurized samples. Thus, plastic formed joints should be structurized with higher hydraulic press pressures to yield highest joint strengths. In case of laser engraved samples, only the optimal value of scanning speed will give highest strength, because both higher and lower speeds facilitate generation of surface structures that are a benefit to joint's strength. In addition, interline distance of engraved samples plays some role in joint strength, and two sided joints are an interesting option because of their strength increase. On the other hand one side connections give the practical balance between moderate strength and short preparation time (no need to replace sample and again laser reprocess).

Final remark is that laser engraving method produces thrice the strong joints when compared to plastic forming, on the other latter method is faster to perform, so both may be useful for consideration.

Acknowledgments. The research was realized at the frame of OPUS grant from Polish National Science Center, contract number 2016/21/B/ST8/03152 and also supported by Ministry of Education and Science, Poland, as a subsidy to Poznan University of Technology (internal subsidy no 0613/SBAD/4710).

Appendix

(See Table 8).

Table 8. Roughness test measurement conditions.

Pick up transducer	
Type	Inductive, diamond stylus
Shape	Pyramid, 90° coning angle
Tip	Rounded, 5 μm radius
Conditions	
Measurement range (vertical)	80 ± 0.04 μm
Cut-off length (lr)	2.5 mm
Evaluation length (ln)	5 mm
Filter	Gauss

References

1. Schmachtenberg, E., Schuck, M., Kuehnert, I.: Assembly injection moulding: forming and assembly in one process. Kunststoffe Int. **04**, 8–16 (2007)
2. Xiping, L., Ningning, G., Can, Y., Shuiping, Z., Shihong, F., Ke, Z.: Aluminum/polypropylene composites produced through injection molding. J. Mater. Process. Technol. **255**, 635–643 (2018)
3. Zhao, S., Kimura, F., Kadoya, S., Kajihara, Y.: Experimental analysis on mechanical interlocking of metal–polymer direct joining. Prec. Eng. **61**, 120–125 (2020)
4. Paul, H., Luke, M., Henning, F.: Combining mechanical interlocking, force fit and direct adhesion in polymer–metal-hybrid structures–evaluation of the deformation and damage behavior. Compos. B Eng. **73**, 158–165 (2015)
5. Xiping, L.F.L., Ningning, G., Can, Y., Bin, W.: Surface topography induced high injection joining strength of polymer–metal composite and fracture mechanism. Comp. Struct. **184**, 545–553 (2018)
6. Honkanen, M., Hoikkanen, M.E., Vippola, M., Vuorinen, J., Lepistö, T.: Metal–plastic adhesion in injection-molded hybrids. J. Adhes. Sci. Technol. **23**, 1747–1761 (2009)
7. Sasaki, H., Kobayashi, I., Sai, S., Hirahara, H., Oishi, Y., Mori, K.: Adhesion of ABS resin to metals treated with triazine trithiol monosodium aqueous solution. J. Adhes. Sci. Technol. **13**, 523–539 (1999)
8. Drummera, D., Schmachtenberg, E., Hülder, G., Meister, S.: MK2—a novel assembly injection molding process for the combination of functional metal surfaces with polymer structures. J. Mat. Process. Technol. **210**, 1852–1857 (2010)
9. Schrickera, K., Stambke, M., Bergmann, J.P., Bräutigama, K., Henckella, P.: Macroscopic surface structures for polymer–metal hybrid joints manufactured by laser based thermal joining. Phys. Procedia **56**, 782–790 (2014)

10. Kim, S.L., Lyu, M.Y.: Adhesive strengths between glass fiber-filled ABS and metal in insert molding with engraved and embossed metal surface treatments. Polym. Eng. Sci. **59**, 96–100 (2019)
11. Bergmann, J.P., Stambke, M.: Potential of laser-manufactured polymer-metal hybrid joints. Phys. Procedia. **39**, 84–91 (2012)
12. Taki, K., Nakamura, S., Takayama, T., Nemoto, A., Ito, H.: Direct joining of a laser-ablated metal surface and polymers by precise injection molding. Microsyst. Technol. **22**(1), 31–38 (2015). https://doi.org/10.1007/s00542-015-2640-2
13. Rodríguez-Vidal, E., Sanza, C., Sorianoa, C., Leunda, J., Verhaegheba, G.: Effect of metal micro-structuring on the mechanical behavior of polymer–metal laser T-joints. J. Mater. Proc. Technol. **229**, 668–677 (2016)
14. Rodríguez-Vidal, E., Sanz, C., Lambarri, J., Quintana, I.: Experimental investigation into metal micro-patterning by laser on polymer–metal hybrid joining. Opt. Laser Technol. **104**, 73–82 (2018). Author, F.: Article title. Journal **2**(5), 99–110 (2016)

Smart Interactive Paints - The Effect of Dispergation on Color Characteristics

Zuzana Tatíčková(✉), Michal Zoubek, Jan Kudláček, Jiří Kuchař, and Viktor Kreibich

Faculty of Mechanical Engineering, Czech Technical University in Prague, Technická 4, 160 00 Prague, Czech Republic
zuzana.fickova@fs.cvut.cz

Abstract. Safety in mechanical engineering is still a topical issue and preventive techniques can prevent workplace injuries. The article focuses on the introduction of a new thermochromic coating system, its development, and its properties. These systems can be used in environments exposed to temperature as an element of active safety. Therefore, if a worker notices a change in color in an exposed area, he will be visually alerted to unsafe contact or a subsequent burn injury. Two series of paints were created with different dispersion methods. The resulting coatings were examined in terms of mechanical properties and especially their color characteristics, which determine the quality and durability of the coating. We verified the safe applicability by long-term monitoring and we can predict the thermochromic behavior of the coating.

Keywords: Thermochromic · Pigment · Interactive · Heat-sensitive · Color · Leuco dye · Safety · CIE lab

1 Thermochromic Coating for Safety Purpose

1.1 Burn Injuries Statistics

Thermochromic coatings aim to propose attention to the increasing temperature of exposed components and visually draw attention to the possible dangers of burn injury. The reaction of the paint is a fundamental change in color shade, which indicates an increase in temperature. Visual stimuli are generally very well perceived, as opposed to reading values from temperature sensors. In addition to the temperature-indicating properties, the newly created thermochromic coating also has good mechanical properties and protects against corrosion.

In mechanical engineering, the risk of burns is relatively high, because heat sources are often used in production. This heat is either direct (molten metal in foundries, heated semi-finished product in forges, heated baths in chemical production) or transferred to individual machine parts, furniture or elements such as handrailings, etc.

According to [3], 1% of the population is treated for burns annually, of which 97% are outpatients and 3% are hospitalized.

B. Gapiński et al. (Eds.): MANUFACTURING 2022, LNME, pp. 175–188, 2022.
https://doi.org/10.1007/978-3-031-00805-4_15

The following graphs were created from pages [4] to evaluate work accidents. These statistics have fixed basic data, always including both sexes (male, female), age (18–60+), and industry only. The category of injury is specified by statistics as "burns, scalds, frostbite", and it is not possible to specify them in terms of individual causes of accidents. The graphs present injuries defined as "serious" and "other" (definition taken from the provider's website). As it is an accident at work, it is almost always covered by an incapacity for work of various durations. However, we cannot rule out that there are minor occupational injuries that are not reflected in the statistics due to nonreporting by the employee (Fig. 1).

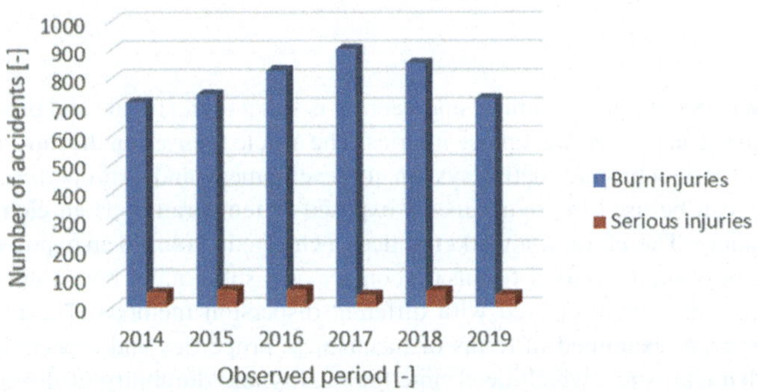

Fig. 1. Total number of accidents and serious injuries in the category "burns, scalds, frostbite" in 2014–2019

1.2 Introduction of Leuco Dye Pigment

According to J.H. Day [5] is defined as "noticeable reversible color change brought about by the boiling point of each liquid, boiling point of a solvent in the case of a solution, or the melting point for solid". The mechanisms of thermochromism are different; here, the mechanism called Leuco dye pigments, which are currently widely used in technical practice will be outlined.

Thermochromic systems based on Leuco dyes are a special system [2], most often in the form of pigment or ink. These composites are colored at room temperature and discolour to an almost transparent state as the temperature increases [14–16].

Opinions on the principle of operation of the leuco dye system are different, despite the fact that this type of thermochromic pigment has been known for a relatively long time [1, 20, 23]. In general, this pigment consists of a chromogenic compound, a developer and a co-solvent. [12, 21, 27].

All three basic components (chromogenic dye, developer, and co-solvent) are represented in the polymer capsule in a specific ratio. In the solid state, the dye reacts with the developer, probably through ion pairs [2] and the open lactone ring of the dye is stabilized by the formation of a chromogenic dye-developer complex [1, 20]. Ring in its colorless form [1] and the developer is dispersed in a liquid solvent [20], leading to negative thermochromia [20, 28] (Fig. 2).

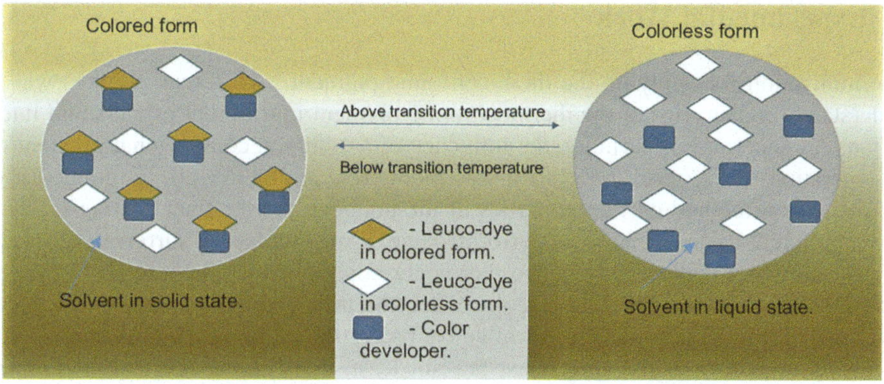

Fig. 2. Operation of a three-component thermochromic system [1]

By chromogenic compound we mean a dye that is sensitive to changes in pH. These are most often based on spirolactones, spiropyrans, fluorates or triphenylmethane derivatives [6, 9, 20]. An important property of these substances is the presence of a lactone ring in the structure, the opening/closing of which causes a color change. When the developer reacts, the lactone ring is opened and a colored zwitterionic structure is formed [2, 19, 27]. Thus, when the lactone ring is closed, it is in a colorless form [9, 12, 21]. The most commonly used chromogenic compound is Crystal violet lactone (CVL) [14, 16, 26]. The chromogenic compound acts as an electron donor [1, 9, 26].

Developers include weak acids, such as phenols, urea - urethane derivatives, gallates [9, 10, 15, 22]. Lauryl gallate is very often used [19, 27] and Bisphenol A (BPA) is preferred for the development of brilliant colors [2, 27]. The developer acts as either a proton donor [12, 15, 20]/electron acceptor [1, 26].

By co-solvent we mean a hydrophobic compound that, by its melting point, directly affects the transition temperature (equivalently also the switching or activation temperature) [9, 10, 22]. Low-melting-point fatty acids, alcohols, and amides are often used [17, 19, 22, 27].

The manufacturers of thermochromic pigments allow the addition of pigments to paints, but in most cases they mention possible problems in the interaction with certain types of base bases, as well as the sensitivity of the pigment to mixing. This information is usually very general [18].

Quite often, it is possible to observe the mixing of thermochromic pigments in exterior facade paints, when the influence of thermochromia on temperature regulation in the vicinity of buildings is investigated [13, 20].

The principle of the actual formulation varies according to the chosen primer and thermochromic medium (ink/pigment). Some authors also add specific types of fillers [1, 10]. In all cases, the pigment is encapsulated to prolong its useful life and reduce possible interactions with the primer.

2 Experimental Work

The motivation for applying the thermochromic pigment to different coatings was to understand the behavior of the thermochromic pigment in different primers and interactions, in general. Furthermore, the influence of a common conventional pigment on the operation of the resulting system was monitored, and above all, the influence of the dispersion method on the properties of the pigment and the final coating material was monitored. Experiments were performed here using high-speed stirrers to verify the functionality and applicability of the thermochromic pigment. The aim was to create a paint that will reversibly change color when the transition temperature is exceeded and that will function as an element of active protection of workers in operations with an increased risk of burn injury. Last but not least, this coating must meet standard conditions such as good applicability by conventional methods, long service life, and corrosion protection. The aim was therefore to produce an interactive paint with high utility properties.

2.1 Color Scheme

At the beginning of the experiment, it was necessary to choose a suitable color scheme. Thermochromic pigments in three colors, yellow, blue, and black, were tested for the zero series. After the conditions were defined, a red thermochromic pigment was selected for further research, mainly because of the appropriate transition temperature. From the experience of the zero series, it was clear that thermochromic pigments do not have sufficient opacity, so a common pigment was added. After reaching the transition temperature, the red pigment discolors, so it was necessary to choose a conventional pigment so that the color change was clearly legible, distinct, and at the same time color-neutral, and did not impair the appearance when used in practice. Thus, the conventional red thermochromic and yellow pigment was chosen, the resulting working scheme – coating at room temperature is orange, at elevated temperature is yellow.

2.2 Materials for Experimental Work

In this work, a pigment based on the Leuco dye system was chosen, a fully reversible pigment of the APEX TH series from the British company Capricorn Specialty Chemicals (currently Chemox Pound). It is an organic microencapsulated material in the form of a colored powder with an average particle size of 7 μm (see Fig. 3). The pigment becomes almost colorless when the transition temperature is reached [18, 25]. A conventional pigment, P.Y. 74 in the form of a pigmented paste.

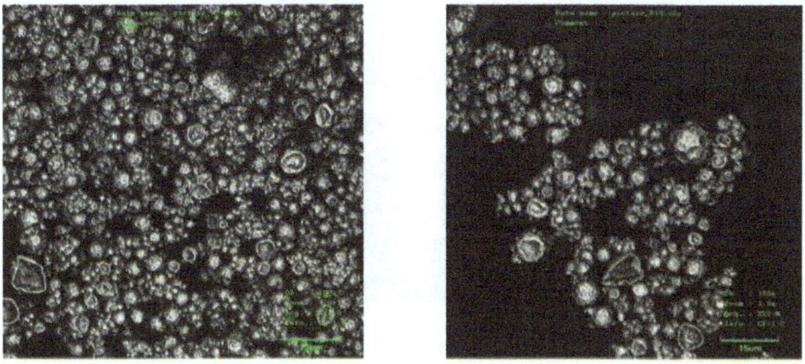

Fig. 3. Images of the thermochromic pigment Apex TH50, from the confocal microscope Olympus Lext OLS 3000

The choice of primer for IP01 was experimental, for easy verification of the basic properties of the thermochromic pigment. For IP02, the same primer was chosen to see how the addition of a conventional pigment would affect the properties of the final coating. The choice of primer for IP03 and IP04 was made through multicriteria decision making.

- Verification of the properties of the already formed formulation, acrylic polyol with yellow pigment and thermochromic pigment. As IP02 [25].
- New paint formulation with self-priming acrylic base, water-soluble, with yellow pigment and thermochromic pigment. As IP03 [25].
- New paint formulation with polyurethane base, with yellow pigment and thermochromic pigment. As IP04 [25].

One of the aims of this work was to verify the applicability of thermochromic pigment under normal production conditions, using high-speed stirrers, and to monitor the effect of different dispersion techniques on the properties of thermochromic paint. These methods were compared.

- Dispersion using saw - tooth stirrer for IP01, IP02, IP03, IP04 (first series)
- Dispersion using rotor-stator stirrer for IP02, IP03, IP04 (second series).

IP01 coating systems were applied to samples of S235JR material to verify the behavior on a metal substrate and determine the mechanical properties (see Fig. 4). The basic mechanical properties of these samples were measured and the behavior of the coating under thermal loading was observed. Acrylic polyol and thermochromic pigment Apex with different transition temperatures (TH 30 – transition temperature 30 °C and TH60 – 60 °C) were used to create the formulation [23, 25].

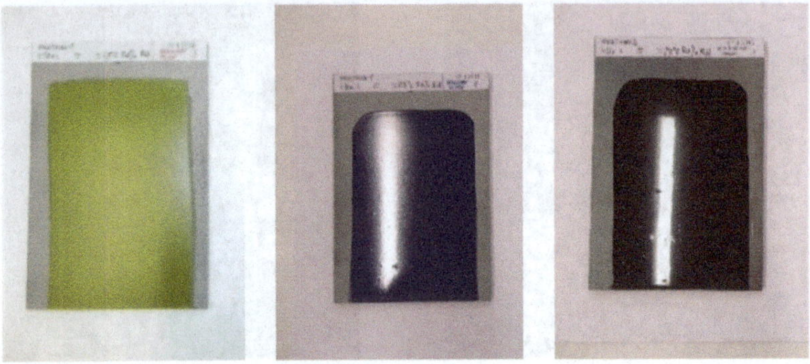

TH30 TH30 TH60

Fig. 4. Samples of the IP01 coating system applied on a metal substrate [25]

Samples of the first series (dispersed by saw - tooth stirrer) IP02, IP03, and IP04 were applied with a film applicator (see Fig. 5), samples of the second series (dispersed by rotor- stator stirrer) IP03, IP04 were created using an airless spraying (see Fig. 6) [24]. In the figures (see Fig. 5, Fig. 6) we can see the different coloration and gloss caused by the dispersion method. Simultaneously, the color change is evident after the transition temperature is exceeded [24].

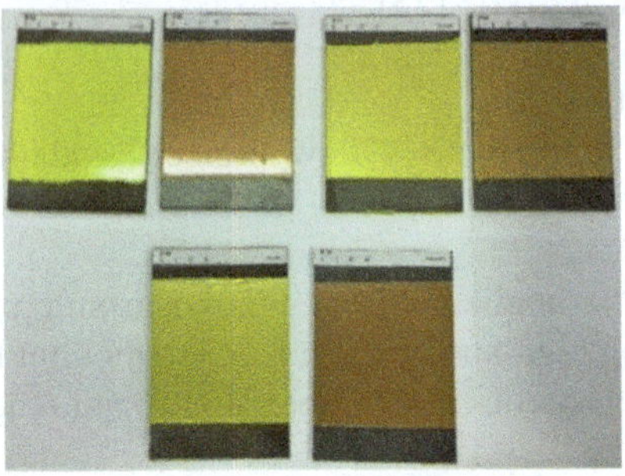

Fig. 5. Results of the first series before testing. Top left: IP02; IP03 and IP04. On the left, the sample is always heated to the color transition temperature [25].

Fig. 6. Results of the second series before testing. From the left IP04 and IP03. On the left, the sample is always heated to the color transition temperature [25].

3 Test Results

The formed samples were subjected to mechanical testing, in particular the adhesion tension, cross-cut test, and the bend test, accelerated test of resistance to UV radiation, corrosion resistance. Results are published in [25]. Color and gloss test result are shown in this chapter.

Color and gloss were measured using a BYK Gardner spectrophotometr. Standard measurements of the IP01, IP02, IP03, and IP04 series were performed on heated samples and samples at room temperature. At a temperature of 20 °C, the curve shows the sample in the original shade and the sample heated over the transition temperature in the so-called discolored state (see Fig. 7). For IP01 samples, the discolored state is characterized by the fact that it "reveals" the basic steel material and the spectral reflectance curve rather characterizes the color of the steel. The measured spectral reflectance curves correspond to the commonly available measured curves of a given color.

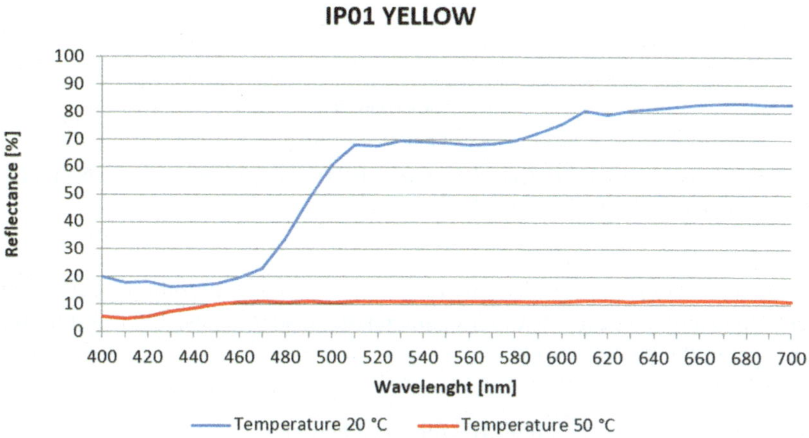

Fig. 7. Spectral reflectance curves of the IP01 sample IP01 yellow at room temperature and after reaching the transition temperature.

For IP02, IP03, and IP04, the graphs of heated samples are very similar in all graphs and correspond to the spectral reflectance curves for yellow. The curve is identical to a paint pigmented only with the base yellow pigment. The curves of thermally unloaded samples are different, they differ according to the specific mixing method and the type of primer.

For a comprehensive overview of the color behavior of the thermochromic pigment, all samples were measured three years apart and placed in a routine laboratory without significant corrosion loading, without increased humidity (up to 60%) and in normal daylight. The aim was to verify the properties during normal "use" of the paint, i.e., whether the legibility of the paint changes more fundamentally.

From the graphs and measured values below, it is clear that the IP02 system shifts the individual color characteristics (see Fig. 8). The sample is significantly lighter in the L* axis, significantly redder in the a* axis, and the difference in the b* axis is clearly noticeable towards yellow. However, in general, the sample retains its color characteristics, so it is possible to safely read the color change on the heated sample.

The IP03 system of the first series changed its characteristics only slightly (see Fig. 9, blue curve). There is a slight difference in the L* axis, the sample is slightly darker. A clearly perceptible difference was noted on the axis b* with a change to yellow. Therefore, the shade after three years is almost identical to the original shade.

For the IP03 sample of the second series (see Fig. 9, green curve), there was a shift in the L* axis, the sample is lighter. A significant difference was observed on the axes a*

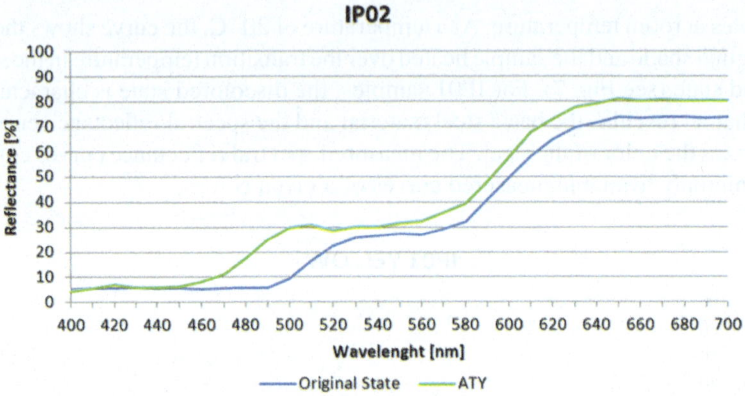

Fig. 8. Comparison of spectral reflectance curves of IP02 sample (at room temperature) in the original condition and after three years (ATY).

and b*, when the sample is significantly redder and yellower after three years. However, in general, the sample retains its color characteristics, so it is possible to safely read the color change on the heated sample.

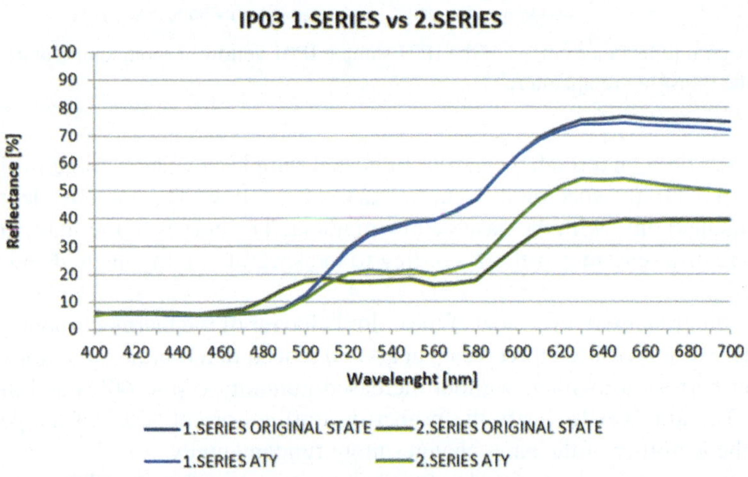

Fig. 9. Comparison of spectral reflectance curves of IP03 samples from the first and second series (at room temperature) in the original state and after three years (ATY).

With the IP04 first series system, there was almost no color shift (see Fig. 10, blue curve). The measurements show that the sample is slightly lighter, slightly yellower, and very slightly redder after three years.

The sample of the IP04 system of the second series (see Fig. 10, green curve) shows a medium difference in the L* axis, the sample is lighter after three years. There is a significant difference and shift in the a* and b* axes, the pattern is greener in the a* axis and bluer in the b* axis. However, in general, the sample retains its color characteristics, so it is possible to safely read the color change on the heated sample.

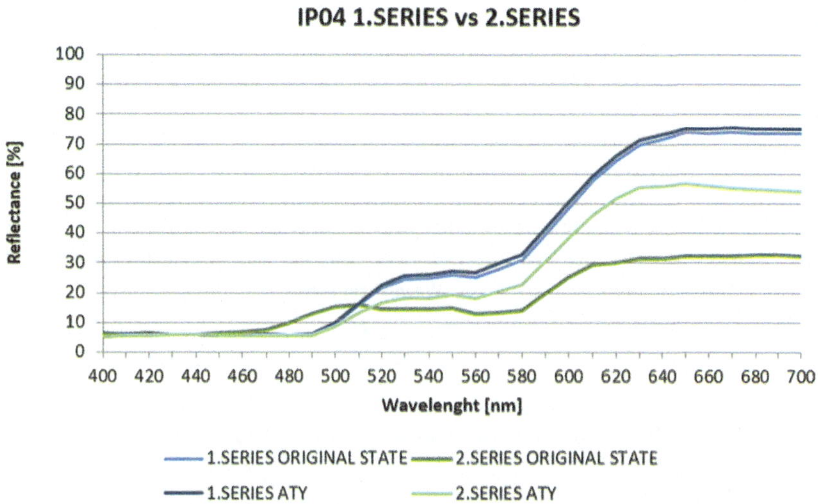

Fig. 10. Comparison of spectral reflectance curves of IP04 samples from the first and second series (at room temperature) in the original state and after three years (ATY).

The comparison of individual shades of the first and second series took place on the original samples and samples after three years. All four curves are plotted in the graphs for clarity. Only systems IP03 and IP04 were compared, because the IP02 second series could not be produced.

When comparing IP03 systems, we can observe a relatively fundamental color difference between the first and second series. From the measured values, it is possible to deduce that the samples have fundamentally different values on the axes L* and b*. The second series sample is therefore darker and bluer than the first series sample. The shift in the a* axis was negligible.

For the IP04 system, we follow the same trend, observing a fundamental shift in all three axes. The 2nd series sample is therefore darker (L*), greener (a*) and bluer (b*).

The difference is also observed in gloss, where the first series samples are brighter, and the second series samples are flat.

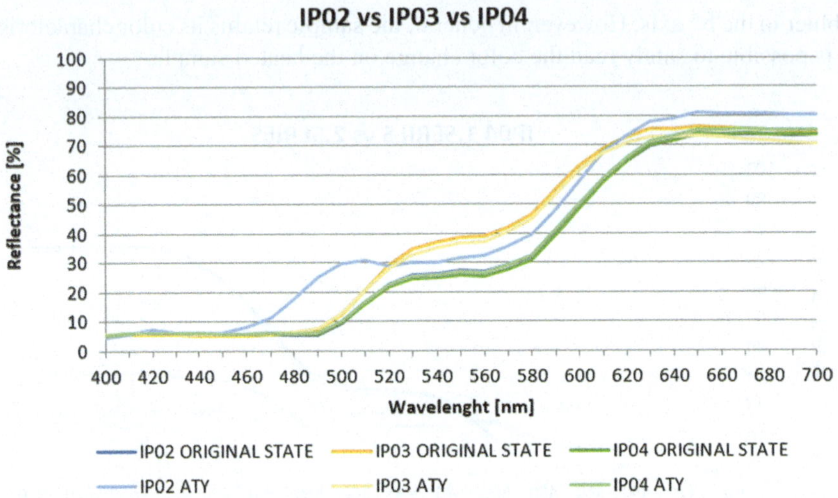

Fig. 11. Comparison of spectral reflectance curves of samples IP02, IP03, IP04 first series (at room temperature) in the original state and after three years (ATY).

When comparing the color shade of the first series samples (see Fig. 11), it is clear that the systems show very good color stability over time, except for IP02. The samples were prepared using the same technological procedure, and despite this, we can observe differences in the color of individual systems. The base of primer therefore also has an effect on the resulting color. The shifts between the individual systems are in all axes. The IP03 system is the lightest, least red, and most yellow. IP02 and IP04 are very

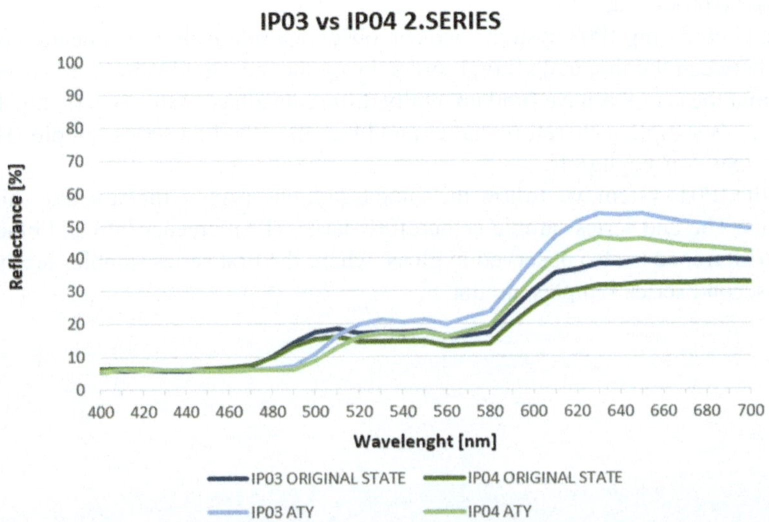

Fig. 12. Comparison of spectral reflectance curves of samples IP03, IP04 second series (at room temperature) in the original state and after three years (ATY).

similar, the largest difference is observed on the b* axis, when IP02 is yellower than IP04.

When comparing the color shade of the second series samples (see Fig. 12), it is clear that they all show a relatively large color shift in time. The samples were prepared by the same technological procedure, the dispersion technology probably negatively affects the properties of the thermochromic pigment.

Table 1 summarizes the gloss measurement, where the effect of the mixing method is obvious. In the first series mixed with a saw - tooth stirrer, the gloss to semigloss appearance of the paint is preserved, and in the second series mixed with the rotor-stator stirrer, the paint is flat.

Table 1. Gloss measurement

Marking	Pretreatment [P – pretreated, NP – no pre-treated]	Median Gloss 60° [GU]	Classification
IP01 Yellow	P	56	Semi-gloss
IP01 Yellow	NP	40	Semi-gloss
IP01 blue	P	88	Gloss
IP01 blue	NP	87	Gloss
IP01 black	P	91	Gloss
IP01 black	NP	90	Gloss
IP02 1.series	P/NP	88	Gloss
IP02 2. series	P/NP	–	–
IP03 1. series	P/NP	70	Gloss
IP03 2. series	P	6,5	Flat
IP04 1. series	P/NP	54	Semi-gloss
IP04 2. series	P	10	Flat

4 Discussion

From the above outputs, is evaluate that the method of dispergation has an effect on color shifts. When mixing with a saw-tooth stirrer, with lower speeds, there is almost no color change in IP03 and IP04, its characteristics are stable.

Mixing rotor-stator devices apparently degrades a certain percentage of pigment particles and causes a more pronounced color shift. However, the fact that the paint of the second series is darker in the original version and significantly brown-orange, despite the color shift, means that there can be no mistake in reading the color in the heated/cold state. It also affects the stability of the hue over time. Paints prepared by this method show a greater color shift in the longer term.

5 Conclusion

In this work, a new thermochromic paint on acrylic polyol (IP02), water-soluble (IP03), and polyurethane (IP04) base was introduced. The IP02 coating system used the same primer as IP01, except that IP02 was further pigmented with the base yellow pigment. This formulation was chosen primarily to verify the properties of IP01. IP02 showed insufficient mechanical properties. Two different dispersion methods have been tested, with the saw tooth disperser and rotor- stator mixer. Experiments have shown satisfactory mechanical properties for both systems IP03 and IP04, with the rotor-stator method appearing slightly more suitable.

When measuring chromaticity after three years with the IP02 system, the individual color characteristics shift on all axes. Despite the relatively fundamental shift, it is possible to safely read the color change on the heated sample.

The IP03 system of the first series changed its characteristics only very slightly, so the shade after three years is almost identical to the original shade.

For the IP03 sample of the second series, there was a larger color shift in all axes. However, in general, the sample retains its color characteristics, so it is possible to safely read the color change on the heated sample.

With the IP04 first series system, there was almost no color shift. The sample of the IP04 second series system shows a significant color shift on all axes, but again the sample retains its color characteristics, so it is possible to safely read the color change on the heated sample.

When comparing the color shade of the first series samples, it is clear that the systems show very good color stability over time, except for IP02. When comparing the color shade of the second series samples, it is clear that they all show a relatively large color shift in time.

Despite the measured color shifts, it is possible in all cases to safely detect a change in color shade, and these paints can easily serve as a safety feature in operation.

From the overall point of view and evaluation of all properties of the formed thermochromic coatings IP02, IP03 and IP04 we can say that the formed coatings have slightly better mechanical properties if the thermochromic pigment is dispersed by the rotor-stator device. However, this method of dispersion degrades a certain part of the pigment and the coating changes its color characteristics over time. The color shifts caused by this are significant, but do not affect the safe use of this system. However, dispersion with a saw - tooth stirrer hardly damages the pigment, the mechanical properties are good and the color shifts are minimal. From this point of view, it is more advantageous to use a conventional saw - tooth stirrer for mixing.

From the point of view of the basic primer for thermochromic coatings, acrylic polyol (IP02) can be clearly excluded from further research, the results of which are unsatisfactory in all respects. The water-based primer (IP03) has very good results in both mechanical and color tests. Its other benefit is the low VOC content and very good application properties. The polyurethane-based primer (IP04) is also of very high quality and can be used even in more demanding corrosion conditions.

The IP03 and IP04 paint formulations have been registered as an utility model [7, 8].

Thermochromic paint formulations are still being investigated. The properties of the samples created with the original paints from the first series are currently being verified. From the preliminary conclusions we can conclude that the coating material IP02 degrades during storage, but the formed coating works to a limited extent. IP03 and IP04 paints do not lose their properties through long-term storage and the created coatings retain their mechanical and color characteristics.

Acknowledgements. Research and experimental verification was carried out within the solution of the project CTU SGS22/OHK2/3T/12 - Influence of surface treatments on the quality of production technologies.

References

1. Aklujkar, P.S., Kandasubramanian, B.: A review of microencapsulated thermochromic coatings for sustainable building applications. J. Coat. Technol. Res. **18**(1), 19–37 (2020). https://doi.org/10.1007/s11998-020-00396-3
2. Bamfield, P., Hutchins, M.G.: Chromic Phenomena: Technological Applications of Colour Chemistry, 2nd edn. Royal Society of Chemistry, Cambridge (2010)
3. Brychta, P., Franců, M.: Vybrané kapitoly z plastické chirurgie a popáleninové medicíny. Brno (2010). https://www.med.muni.cz/Traumatologie/Popaleniny/Popaleniny.htm
4. BOZP v číslech a grafech. BOZP Statistiky. Praha (2021). http://statistikybozp.vubp.cz/
5. Day, J.H.: Thermochromism. Chem. Rev. **63**(1), 65–80 (1963). https://doi.org/10.1021/cr60221a005
6. Dawson, T.L.: Changing colors: now you see them, now you don't. Color. Technol. **126**, 177–188 (2010). https://doi.org/10.1111/j.1478-4408.2010.00247.x
7. Drbout, V., Háková, M., Kudláček, J., Tatíčková, Z., Kreibich, V.: Nová termochromická nátěrová hmota s akrylátovým základem. Functional Sample (2019)
8. Drbout, V., Háková, M., Kudláček, J., Tatíčková, Z., Kreibich, V.: Nová termochromická nátěrová hmota s polyuretanovým základem. Functional Sample (2019)
9. Hajzeri, M., Bašnec, K., Bele, M., Gunde Klanjšek, M.: Influence of developer on structural, optical and thermal properties of a benzofluoran-based thermochromic composite. Dyes Pigm. **113**, 754–762 (2015). https://doi.org/10.1016/j.dyepig.2014.10.014
10. Karlessi, T., Santamouris, M., Apostolakis, K., Synnefa, A., Livada, I.: Development and testing of thermochromic coatings for buildings and urban structures. Sol. Energy **83**(4), 538–551 (2009). https://doi.org/10.1016/j.solener.2008.10.005
11. Kim, I.J., Ramalingam, M., Son, Y.-A.: Investigation of reversible self-thermochromism in microencapsulated fluoran-based materials. Dyes Pigm. **151**, 64–74 (2018). https://doi.org/10.1016/j.dyepig.2017.12.047
12. Li, Y., Wang, Q., Zheng, X., Li, Y., Luan, J.: Microcapsule encapsulated with leuco dye as a visual sensor for concrete damage indication via color variation. RSC Adv. **10**(3), 1226–1231 (2020). https://doi.org/10.1039/C9RA09492J
13. Ma, Y., Zhu, B., Wu, K.: Preparation of reversible thermochromic building coatings and their properties. J. Coat. Technol. **72**, 67–71 (2000). https://doi.org/10.1007/BF02720527
14. MacLaren, D., White, M.A.: Design rules for reversible thermochromic mixtures. J. Mater. Sci. **40**(3), 669–676 (2005). https://doi.org/10.1007/s10853-005-6305-x
15. Özkayalar, S., Adigüzel, E., Aksoy, S.A., Alkan, C.: Reversible color-changing and thermal-energy storing nanocapsules of three-component thermochromic dyes. Mater. Chem. Phys. **252**, 12 (2020). https://doi.org/10.1016/j.matchemphys.2020.123162

16. Panák, O., Držková, M., Kaplanová, M.: Insight into the evaluation of colour changes of leuco dye based thermochromic systems as a function of temperature. Dyes Pigm. **120**, 279–287 (2015). https://doi.org/10.1016/j.dyepig.2015.04.022
17. Panák, O., Držková, M., Kaplanová, M., Novak, U., Gunde Klanjšek, M.: The relation between colour and structural changes in thermochromic systems comprising crystal violet lactone, bisphenol A, and tetradecanol. Dyes Pigm. **136**, 382–389 (2017). https://doi.org/10.1016/j.dyepig.2016.08.050
18. Product information list: Apex Pigments. Littleport (2014)
19. Seeboth, A., Lötzsch, D.: Thermochromic and thermotropic materials. Pan Stanford, Singapore (2014)
20. Zheng, S., Xu, Y., Shen, Q., Yang, H.: Preparation of thermochromic coatings and their energy saving analysis. Sol. Energy **112**, 263–271 (2015). https://doi.org/10.1016/j.solener.2014.09.049
21. Rossi, S., Simeoni, M., Quaranta, A.: Behavior of chromogenic pigments and influence of binder in organic smart coatings. Dyes Pigm. **184**, 19 (2021). https://doi.org/10.1016/j.dyepig.2020.108879
22. Tang, H., MacLaren, D., White, M.A.: New insights concerning the mechanism of reversible thermochromic mixtures. Can. J. Chem. **88**(11), 1063–1070 (2010). https://doi.org/10.1139/V10-069
23. Tatíčková, Z., Kudláček, J., Heller, J., Matas, F., Pepelnjak, T.: Interactive painting systems. In: IN-TECH 2015 - International Conference on Innovative Technologies, pp. 447–450. Faculty of Engineering University of Rijeka, Rijeka (2015)
24. Tatíčková, Z., Kudláček, J., Heller, J.: Interactive thermosensitive painting systems for safety purposes. In: IN-TECH 2018 International Conference on Innovative Technologies, pp. 227–230. Faculty of Engineering University of Rijeka, Zagreb (2018)
25. Tatíčková, Z., Kreibich, V., Kudláček, J., Svoboda, J.: Thermochromic painting systems – a new approach to increase occupational safety. In: METAL 2021 30th Anniversary International Conference on Metallurgy and Materials, vol. 30, pp. 593–599. Tanger, Brno (2021). https://doi.org/10.37904/metal.2021.4187
26. Zhu, C.F., Wu, A.B.: Studies on the synthesis and thermochromic properties of crystal violet lactone and its reversible thermochromic complexes. Thermochim. Acta **425**(1–2), 7–12 (2005). https://doi.org/10.1016/j.tca.2003.08.001
27. Vik, M., Periyasamy, A.P.: Chromic Materials: Fundaments, Measurements, and Applications. Apple Academic Press, Canada (2018)

Technology of Repairs of Anticorrosion Protection of Smaller Parts of Historic Steel Structures Using Hot-Dip Galvanizing

Jan Kudláček[1(✉)], Jakub Svoboda[1], Michal Zoubek[1], Henryk Kania[2], and Mariola Saternus[2]

[1] Faculty of Mechanical Engineering, Czech Technical University in Prague, Prague, Czech Republic
jan.kudlacek@fs.cvut.cz

[2] Faculty of Materials Engineering, Silesian University of Technology, Katowice, Poland

Abstract. Protection of steel structures against corrosion is very important to ensure proper function and prolong service life. Many standards and procedures have been developed for the correct solution of anti-corrosion treatments. External steel structures stressed by forces and atmospheric influences are of the greatest importance. This paper focuses on the design of suitable corrosion protection for smaller parts of historic or listed steel structures respecting the current conditions of the corrosive environment to which these buildings are exposed.

Keywords: Corrosion protection · Hot dip galvanizing · Duplex system · Steel construction

1 Corrosion Protection of Steel Structures

Steel structures, such as bridges, high-voltage pylons, railings and the like, are subject to environmental influences. The most commonly used materials for these structures are steels of the usual strengths. These materials are subject to corrosion in the particular environment. Corrosion attack of steel structures occurs mainly due to the loss of anti-corrosion protection of the selected anti-corrosion system. Corrosion can lead to a gradual weakening of the steel structure, reduction in its static capacity, increase in dynamic stress, or even a potential loss of safety of the entire structure. The estimated service life of corrosion protection (CP) should be around 15–30 years. The most common forms of corrosion attack in historic steel structures are uniform, point, pitting and crevice corrosion. For these reasons, it is necessary to choose a suitable anti-corrosion system with regard to the current environmental conditions, which have a major impact on the life of anti-corrosion protection and which were not taken into account during the design and implementation time. For the correct choice of appropriate corrosion protection of steel structures, it is always necessary to consider the environment to which the structure is exposed, the design of the steel structure, protection capability of the particular CP, technological aspects of CP implementation with respect to the environment and last but not least the planned service life of CP and the economic aspect [1].

B. Gapiński et al. (Eds.): MANUFACTURING 2022, LNME, pp. 189–201, 2022.
https://doi.org/10.1007/978-3-031-00805-4_16

The paper deals with research and development of a proven technology for repairs of corrosion protection of disassemble parts of steel structures, mainly dynamically stressed parts. The paper is structured in describing the history of corrosion protection, as it was done at the beginning of the 20th century, verifying the suitability of the base material for the application of hot dip galvanizing (using spectral analysis SEM). Determining the appropriate surface pretreatment technology before the application of corrosion protection (method of removing the original corrosion protection and ensuring the most optimal surface pretreatment with respect to the surface structure, the effect of abrasive blasting, thermal stripping and their combination) was investigated. The main output is the design of the technological process and the method of application of corrosion protection, reflecting the current requirements for the functionality and durability of the design of corrosion protection.

2 History of Corrosion Protection of Steel Structures

Corrosion protection of steel structures (riveted bridge structures) in the period of their creation was always solved only by coatings. Electrochemical protection (cathodic protection), for example of steel pillars, was used except for welded bridge structures. In general, it can be stated that the method of corrosion protection by organic coatings with pigmented corrosion inhibitors was used, which was widely extended at that time. As for the application of organic coatings, in particular flax oils (vermilion; practically the only option in the middle of the 19th century) pigmented with red lead (Pb_3O_4) were used at the beginning. Red lead was a very effective corrosion inhibitor used in a significant mixing ratio to primers almost until the middle of the 20th century. Organic coatings pigmented with red lead (Pb_3O_4) or other lead compounds, such as dicalcium lead ($2CaO \cdot PbO_2$, Ca_2PbO_4), or lead cyanamide or lead benzoate were used as a corrosion protection on most steel structures in Europe. However, it should be noted that in the case of the use of lead compounds, red lead clearly dominated. Red lead pigmented paints should not be applied as topcoats due to the fact that red lead is unstable due to the action of air CO_2, H_2S, SO_2 and also because it degrades as a result of exposure to UV radiation [2]. Therefore, it has always been recommended to apply the top coats of the coating system immediately after the red lead pigmented primer has dried.

Corrosion protection of steel structures built on the territory of today's Czech Republic in the 19th century was carried out with oil paints. These paints were based on purely natural and subsequently also partially technologically modified drying oils and were widely used as a common protection of steel elements against atmospheric corrosion. Of course, they were also used as top coats pigmented with ground limestone, float chalk, kaolin, carbon black, ground gypsum, talc. Colored clays (especially ferrous), asphalt fractions and tars were also used for pigmentation. Powders of finely ground brown and black coal have also been used. Less often, synthetic inorganic pigments based on Pb, Cr, Zn, etc. were used [3].

3 Repairs of Corrosion Protection of Steel Structures

The issue of corrosion protection of steel structures is extensive and its specific methods are based primarily on the level of knowledge of workers who deal with it. There is no single rule for its implementation. The specific procedure for repairing corrosion defects should be performed by an authorized person with appropriate qualifications. He should be able to determine which corrosion defect it is and what technology to prescribe for repair and maintenance.

Various defects and damage can occur on steel structures. These defects can generally include corrosion, mechanical damage, improper steel structure design, corrosion protection (CP) system defects, vandalism, and more. The concept of corrosion protection must include: the material and its protection for the given technology, data on material losses and its monitoring during operation of the equipment. Moreover, the operational reliability, durability, fire resistance and economy of the solution must be taken into consideration as well. This whole part is important and often determines the success of the whole repair investment. At the design stage of the project repair, special care should be taken to ensure adequate service life of the structure.

The basis of the CP repair proposal is a realistic assessment of defects and the causes of their occurrence, which in most cases is the result of a routine or main inspection of the steel structure. When preparing a draft technological regulation (TR) for CP repair, the basic data of the steel structure (material, existing CP system, state of the environment, structure, etc.) and the suitability of using possible repair technology for the selected type (local repair, partial repair, renewal of the whole CP system) are always used. When drafting a technological regulation, it is necessary to determine suitable partial technological processes that will lead to ensuring the required service life of CP. The most important process is surface pretreatment technology. It is always necessary to pre-treat the surface so that the subsequent technological process (paint, thermal spraying, thermal plating) has sufficient adhesion to the original surface and prevents the occurrence of subsequent defects in corrosion protection. The draft of technological regulations can be based on the regulation ŘSD TKP 19 B and C or the regulations of the Railway Administration and EN ISO 12944 [4, 5].

4 Repair of the Railing of the Listed Čech Bridge

Čech Bridge is the shortest of the bridges leading across the Vltava in Prague. It is a continuation of today's Pařížská Street and connects Dvořák's embankment on the Old Town side with Edvard Beneš's embankment on the Letná side. The bridge is 169 m long and 16 m wide, used for trams, road transport and pedestrians. The Art Nouveau bridge was put into operation in 1908. The bridge has a rich decoration and is a protected technical monument. The bridge was designed by Jiří Soukup, František Mencl and Václav Trča, and the author of the architectural design was prof. Jan Koula. After the bridge was put into operation in 1908, it was named the Svatopluk Čech Bridge. From 1940 to 1945, during the occupation, it was renamed Mendel Bridge. Since 1945 it has been called Čechův most (Čech Bridge) [6] (Fig. 1).

Fig. 1. Historical photograph of Čech Bridge [7]

4.1 Initial State of Corrosion Protection of the Čech Bridge Railing Segments

As part of a local investigation, it was found that the railing is exposed to an adverse corrosive environment. According to ČSN EN ISO 12944-2, this environment can be characterized as environment C4 - significant atmospheric corrosion stress (in winter also due to the effect of spreading materials) [8]. Furthermore, a large number of deficiencies affecting the protective properties of CP and the service life of the steel structure were found. These are mainly the following defects and shortcomings:

- Corrosion of the column feet and thus irreversible degradation of materials in some columns.
- Subcorroding and corroding of the coating system - inappropriate and ineffective existing CP.
- Mechanical damage to CP - caused by operation/traffic or vandalism.
- Inhomogeneous coating system.
- Slit corrosion on riveted steel railing elements. The structure is subjected to dynamic loading of steel elements and due to this effect, the existing CP cracks in the structural joints with the subsequent corrosion attack of the base material.
- Galvanic corrosion attack due to improper connection of different materials.
- Last but not least, neglected maintenance of the structure, dirt on the upper surfaces of the supporting structure.
- Uneven coating system of the railing CP.
- Corrosion of feet and welded joints of the railing.

In the past, corrosion protection was applied to partial railing elements of Čech Bridge, using a lead-based primer and several coats of a paint system, which no longer fulfilled its required anti-corrosion function. As a result, an irreversible corrosion of some parts of the railing occurred and the life and functionality of the corrosion protection was lost. For this reason, the repair of the railing was started with the overall restoration of corrosion protection (Fig. 2).

Fig. 2. Current state of corrosion protection of the Čech Bridge railing

5 Design of Corrosion Protection of Railing

As part of corrosion protection, the restoration of CP was originally suggested and tested with a multilayer coating system with a Zn-based primer. However, since it is a forged riveted railing with the existence of a large number of cracks, which make the proper surface protection difficult, corrosion attacked the test segments only after a year of exposure in the given environment and thus the selected CP system lost its function.

For this reason, and based on the findings, it was recommended to try to apply on particular segments of the railing an anti-corrosion system using hot-dip galvanizing. The combination of hot-dip galvanized coating, which will provide anti-corrosion protection by barrier and especially cathodic principle in case of damage to CP and a subsequent multilayer coating system (duplex system). Duplex coating system comprises of a layer from hot dip galvanizing and a paint coat [9, 10].

Moreover, the metallic coatings protect steel sacrificially at damaged areas or at small pores in the coatings [11]. The durability of the duplex coating in a corrosive environment is, therefore, very dependent on the performance of the paint coating [12].

Proposed type of corrosion protection of Čech Bridge railing: hot-dip galvanizing + ONS 93 (registration number ONS 007/2020 according to S5/4) [13, 14].

1. Hot dip galvanizing 100 μm
2. PPG SigmaFast 278 primer - cream shade 160 μm
3. PPG SigmaFast 210 HS top coat - RAL 7044 80 μm

Total NDFT 340 μm.

5.1 Verification of the Suitability of Corrosion Protection of the Railing

For the suitability of the base material for hot-dip galvanizing, the material composition was determined. Three samples of partial elements were determined and taken from a selected segment of the Čech Bridge railing. The locations of the samples taken are shown in Fig. 3.

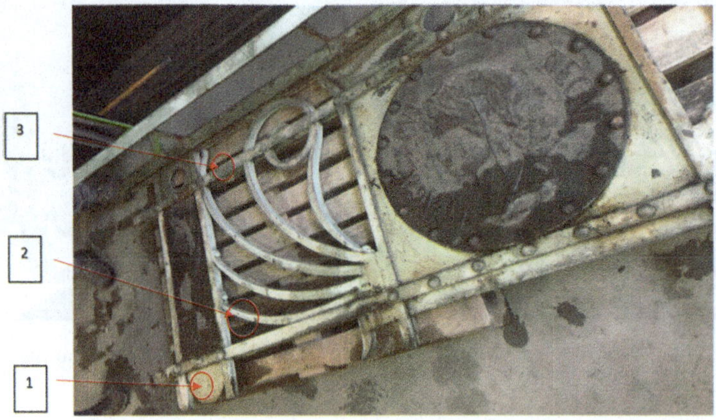

Fig. 3. Locations of samples taken to determine the composition of the material

From the material analysis it was found that samples 2 and 3 are - according to their composition - suitable for hot dip galvanizing. According to the determined material composition, sample 1 is not suitable for hot-dip galvanizing. The reason is its unsuitable Si content in the base material. According to the performed measurement, the Si content in sample 1 is 0.086%, which corresponds to the Sandelin range, where materials with a silicon content of 0.04 to 0.14% achieve uneven and thick coatings [15, 16]. Despite the finding that some elements of the railing segment are not suitable for hot-dip galvanizing, it was decided to perform test galvanizing on two railing segments. In order to be able to hot-dip galvanize the segments, the original paints had to be removed. This removal in the first phase was carried out by blasting with a sharp-edged blasting agent. Figure 4 shows a blasted surface with remnants of the original paint.

Fig. 4. Pre-treatment of the surface by blasting before application of hot-dip zinc by immersion

Hot-dip galvanizing and verification of the suitability of the application of hot-dip galvanizing on the railing segments of Čech Bridge was performed in the hot-dip galvanizing plant. After the application of hot-dip galvanizing in places between the overlapped materials and in hard-to-reach parts (crevices and inner corners), where 100% cleaning did not occur and the original paint remained, there were undesirable defects caused by hot-dip galvanizing (non-plated places and inconsistent hot-dip zinc coating in places with burnt paint).

To verify 100% surface cleaning, it was necessary to perform additional dezincification on one railing segment and thus remove the remaining burnt paint. Figure 5 shows

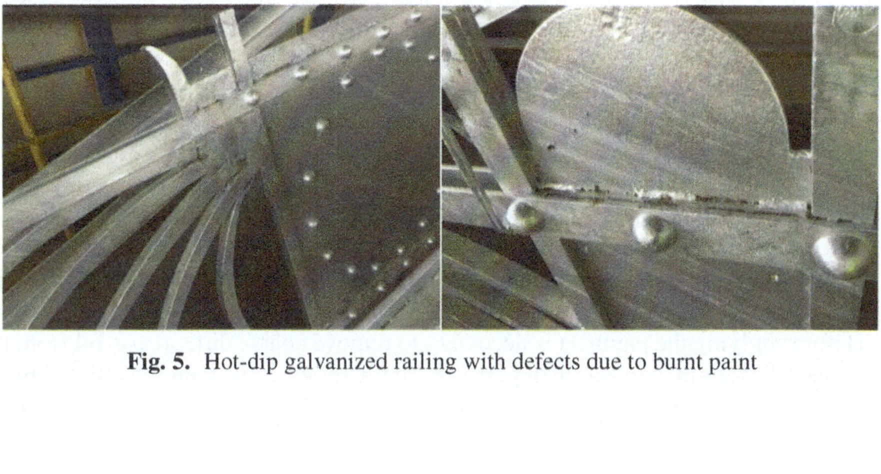

Fig. 5. Hot-dip galvanized railing with defects due to burnt paint

a hot-dip galvanized railing with defects due to burnt paint and Fig. 6 shows a hot-dip galvanized railing without defects due to burnt paint.

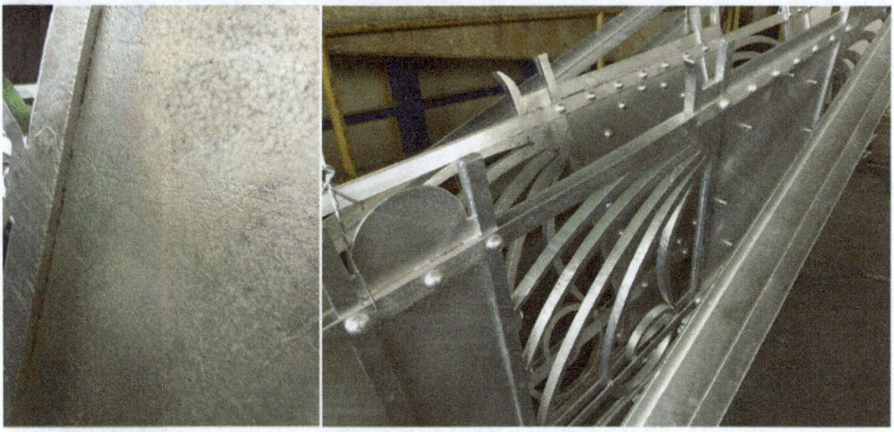

Fig. 6. Hot-dip galvanized railing without defects due to burnt paint

Due to the need for 100% cleaning and removal of the original paints, a better cleaning of the railing segments was applied in the second phase. A combination of thermal paint stripping followed by a blasting to a surface cleanliness of Sa 3 grade. Such a pretreated surface was already suitable for hot-dip galvanizing without numerous defects, see Fig. 7.

5.2 Technological Process of Corrosion Protection of Čech Bridge Railing Segments

1. Checking the condition of the bridge railing surface (checking the blacksmith work, the surface must be free of scale, spatter, welds reground without cracks, edges rounded to R = min. 2 mm, paint residues must be removed mechanically, surface inspection after removal of the original paints and the resulting corrosion attack - after thermal paint removal and subsequent blasting to a surface cleanliness of Sa 3 grade [17]).
2. Hot-dip galvanizing in accordance with ČSN EN ISO 1461, EN ISO 14713-1, EN ISO 14713-2 (Pickling and hot-dip galvanizing of the steel railing construction) (Figs. 8 and 9).
   Unacceptable defects in the hot-dip galvanized coating: insufficient coating thickness, discontinuous areas of the coating, coating impurities (ash, flux residues, etc.), defects in the coating resulting from the steel substrate (scales, weld defects, etc.). The defect is not the gloss or matte surface of the coating or white rust. Also, in the riveted joints, light residues of the original colors are allowed.
3. Before applying the paints, it is necessary to remove coarse dirt, grease, oil from the surface of hot-dip zinc and then perform a light abrasive sweeping to remove white rust from the surface and at the same time to allow the paint to anchor. The sweeping

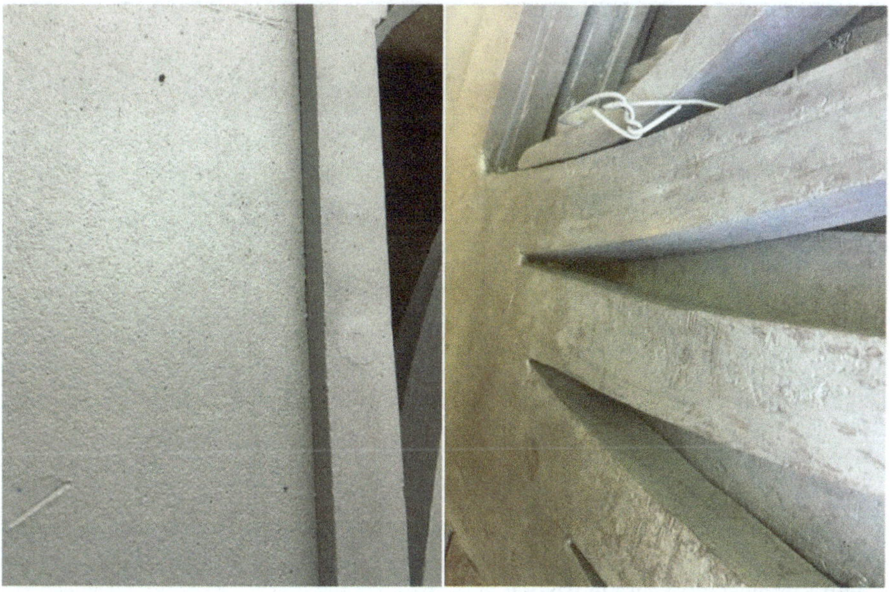

Fig. 7. Pre-treatment of the surface by blasting before application of hot-dip zinc by immersion - surface cleanliness of Sa 3 grade

Fig. 8. Pickling before application of hot-dip zinc by immersion

shall be carried out using a non-metallic blasting medium - corundum, fraction 0.2–0.5 mm, at pressure of 2.5 bar, with nozzle diameter of 8 mm, nozzle distance from the surface of approx. 400 to 500 mm, at angle of 30–60 °C, with an expected zinc loss of max. 5–10 μm - the residual thickness of the hot-dip zinc coating after

Fig. 9. Demonstration of hot dip galvanizing application.

sweeping must be checked by non-destructive measurement in accordance with the ČSN EN ISO 2178 standard. The roughness check can be performed in accordance with Rugotest No.3, grade BN-9a.

4. Climatic conditions are checked before applying the primer coat of the selected corrosion protection system. The control of climatic conditions for the implementation of CP must be in accordance with the ČSN EN ISO 8502-4 standard. The temperature of the structure must be at least +3 °C higher than dew point temperature. The minimum air temperature when applying CP must be +5 °C. The maximum permissible relative humidity when applying CP is 75%.
   Furthermore, before applying the primer, the surface must be checked for the presence of dust particles - the amount and size of dust particles shall be max. 2-2 according to ČSN EN ISO 8502-3. If the surface is contaminated, the dirt is removed with an industrial vacuum cleaner, or with compressed air - dry, free of greasy dirt. Then another inspection is performed.
5. Within 4 h after sweeping, a primer is applied according to the selected CP system. Problematic areas (riveted parts, welds, corners and edges) are first treated with a stripe coat. Only after the flash time of this coating, the remaining surfaces (including already painted problematic areas) are treated to the required coating thickness according to TR by cross-spraying. In case of unsatisfactory coating thicknesses, the thickness is supplemented by another spray. Prior to the application of the intermediate layer and the top coat, stripe coat is applied to all problematic areas (riveted parts, welds, corners and edges). The application of stripe coatings will be followed by the application of intermediate layers and a top coat.
   Unacceptable defects of the coating system - insufficient thickness, unpainted areas, wrinkles, shrinkage, blisters, peeling (delamination), cracks, craters. Defects in the coating system must be repaired by mechanical regrinding with a smooth and gradual transition to a perfect surrounding coating and by applying the paint to form a coating of the required thickness.

Figure 10 shows the application of the overall corrosion protection to the Čech Bridge railing segment, which will meet the required properties and guarantee a long service life in a highly corrosive environment.

Fig. 10. Application of the overall system of corrosion protection to the Čech Bridge railing segment

The proven technology is fully usable in the areas of design and application of CP of disassemble parts of steel structures and bridges and its use should be required by the monument care authorities after the designers (always in agreement with the corrosion engineer). The scope of use should correspond to the monumental significance of the structure. The proven technology can also be used for the work of monument care authorities, which will thus receive instructions for the implementation of suitable CP during the repair, reconstruction or reconstruction of monuments.

6 Conclusion

As part of the repair of the railing on Čech Bridge, it was possible to apply a unique technology for the renewal of corrosion protection to smaller parts of the historic steel structure using hot-dip galvanizing. This technology has been approved by the Institute of Monument Care and will guarantee the required service life of corrosion protection (more than 25 years) in a C4 corrosive environment (corrosive aggressiveness of the atmosphere). The coating system was selected from a system of approved protective coating systems. A coating system from the renowned paint manufacturer PPG. The aim was to apply such a corrosion protection, which will provide protection even on very exposed and mechanically stressed parts of the railing, where there could be a rapid corrosion attack and loss of CP life.

A total of 100 partial segments of railings with stands and other decorative elements will be repaired on Čech Bridge in the manner of the proven proven technology. The most decorative elements treated in the forced way are the stands for anchoring the railing segments. The overall repair and renewal of the CP will take 2 years, this is due to the restoration modifications of the corrosively attacked elements of the Čech Bridge railings. The main scientific parameter is to increase the service life of CP for the given corrosive environment in which the steel and bridge structures are located. Using proven technology, the service life of CP should increase by at least 100% to at least 25 years compared to conventional CP, which is planned to last for 5 to 10 years. There are also financial savings associated with this.

Acknowledgements. Research and experimental verification was carried out within the solution of the project for the Ministry of Culture of the Czech Republic - NAKI DG18P020VV033 - Methods for ensuring the sustainability of steel bridge structures of industrial cultural heritage.

References

1. Beitelman, A.: Maintenance painting of steel structures. Technical manuscript M-189 (1976)
2. Kreislova, K., Knotkova, D.: The results of 45 years of atmospheric corrosion study in the Czech Republic. Materials **10**(4) (2017). ISSN 1996-1944. https://doi.org/10.3390/ma1004 0394
3. Brokbartold, M., Grupe, M., Marschner, B.: Effectiveness of different soil amendments to reduce the Pb and Zn extractability and plant uptake in soils contaminated by anticorrosion paints beneath pylons. J. Plant Nutr. Soil Sci. **175**(3), 443–455 (2012). https://doi.org/10. 1002/jpln.201100198
4. Technical qualitative conditions for building of land communications: Corrosion protection of steel bridges and structures, Chapter 19B. Fourth wording. Ministry of Transport, Department of Road Infrastructure, Prague (2018)
5. Technical qualitative conditions for building of land communications: Corrosion protection of steel bridges and structures during repairs and reconstructions, Chapter 19C. Ministry of Transport, Department of Land Communications, Prague (2021)
6. Kudláček, J., Pokorný, P., Pokorný, R.: Experimental verification of blasting and corrosion removal methods on the bridge under Vyšehrad. Final Report. CTU in Prague, Faculty of Civil Engineering, Prague (2019)
7. http://www.fotohistorie.cz/Praha/Praha-mesto/mosty/Cechuv_most/Default.aspx
8. EN ISO 9223. Corrosion of metals and alloys: corrosivity of atmospheres. Classification (2012)
9. Technical standard TN33. Coating and remediation of steel structures of power lines and power stations. Revision 6. ČEPS (2015)
10. Siostrzonek, R., et al.: Corrosion protection of structural steelwork by a duplex coating system. Materials Science Forum 818. Trans Tech Publications Ltd. (2015). https://doi.org/10.4028/ www.scientific.net/MSF.818.162
11. Price, S.J., Figueira, R.B.: Corrosion protection systems and fatigue corrosion in offshore wind structures: current status and future perspectives. Coatings **7**(2), 25 (2017). https://doi. org/10.3390/coatings7020025

12. Knudsen, O.Ø., et al.: Experiences with thermal spray zinc duplex coatings on road bridges. Coatings **9**(6), 371 (2019). https://doi.org/10.3390/coatings9060371
13. SŽDC S5/4. SŽDC S5/4: Corrosion protection of steel structures. Správa železniční dopravní cesty, státní organizace (Railway administration, state organization) (2019)
14. EN ISO 12944-3. Paints and varnishes - Corrosion protection of steel structures by protective paint systems: Part 3: Design considerations (2018)
15. Kuklík, V., Kudláček, J.: Hot-Dip Galvanizing of Steel Structures. Butterworth-Heinemann (2016)
16. Kania, H., Saternus, M., Kudláček, J.: Impact of Bi and Sn on microstructure and corrosion resistance of Zinc coatings obtained in Zn-AlNi bath. Materials **13**(17) (2020). ISSN 1996-1944. https://doi.org/10.3390/ma13173788
17. EN ISO 8501-1. Preparation of steel substrates before application of paints and related products: Rust grades and preparation grades of uncoated steel substrates and of steel substrates after overall removal of previous coatings (2007)

A Comparative Experimental Study of Acids Used for Anodic Oxidation for the Purposes of Aerospace Industry

Eva Michelle Sedláčková(✉), Jiří Kuchař, Viktor Kreibich, and Jakub Horník

Faculty of Mechanical Engineering, Czech Technical University in Prague, Prague, Czech Republic
evamichelle.sedlackova@fs.cvut.cz

Abstract. The article compares 4 types of anodic oxidation technology in terms of corrosion resistance, ecological and technological point of view. The work compares two technologies of anodic oxidation (chromic acid, sulfuric acid), which have been used in practice for an extensive period of time, and two newer technologies (tartaric acid, boric acid) in order to replace the above-mentioned chromic acid due to its toxicity. The experimental part of this article uses a corrosion chamber, weight test and coating adhesion test - grid test to evaluate qualitative parameters of the created protective layers.

Keywords: Anodic oxidation · Corrosion · Aerospace industry · Aluminum alloys · Tartaric acid · Boric acid

1 Introduction

Several types of surface treatment technologies are used in the aerospace industry. This article deals with the anodic oxidation of aluminum alloys and the types of acids used in this technology. The best-known acid used in anodic oxidation is sulfuric acid, which has very good corrosion resistance, it can also be painted with pigments to change the color surface according to our specifications. This is very advantageous for parts that serve as a way to properly discern the aircraft. Another very common acid is chromic acid. The layer formed by this method shows very high corrosion resistance and excellent surface adhesion. The main disadvantage is the content of hexavalent chromium, which is very toxic and carcinogenic [1]. The European Union is therefore seeking to limit this chemical through the regulation of REACH (a European Union regulation aimed at improving the protection of human health and the environment) [2].

In succession to this problem, the method of anodic oxidation using tartaric acid and a mixture of boric acid and sulfuric acid was introduced into practice. The experimental part compares the above-mentioned methods of anodic oxidation with anodic oxidation (AO) using chromic acid and sulfuric acid in terms of maximum corrosion resistance and in economic terms [1, 3, 4].

B. Gapiński et al. (Eds.): MANUFACTURING 2022, LNME, pp. 202–215, 2022.
https://doi.org/10.1007/978-3-031-00805-4_17

2 Anodic Oxidation

Anodic oxidation or anodizing is one of the most common surface treatments of aluminum and its alloys. This technology is very often used in the aerospace industry. It is an electrolytic process in which a layer of aluminum Al2O3 is formed using suitable baths. The overall layer consists of a non-porous, dielectric insulating layer and a finely porous cover layer, which is formed from the insulating layer when redissolved in the electrolyte [1, 5].

The whole process of anodic oxidation is affected by certain conditions, which give us the properties of the resulting layer. These conditions include type of electrolyte, bath temperature and current density. The process parameters are further linked to the type of anodic oxidation, which is divided as decorative oxidation and hard oxidation [5].

Decorative anodic oxidation serves primarily to improve the appearance of the surface and to improve corrosion resistance. Hard anodic oxidation serves to increase functional properties [1, 5].

The main reason for using this technology is to increase corrosion resistance, abrasion resistance and increase hardness. Furthermore, by selecting suitable parameters (decorative AO), the components can be colored very easily. However, its disadvantages include its non-ecological nature. As these are very strong acids that are dangerous for the environment and the operator [1, 5].

3 Experiment Description

Four acids for anodic oxidation were selected for the experiment, namely chromic, sulfuric, boric, and tartaric acid. Sulfuric acid is the best-known acid used in anodic oxidation, and chromic acid is a very common acid. This acid contains hexavalent chromium, which is highly toxic, and efforts are being made to reduce the usage of this chemical. For this reason, anodic oxidation in boric acid and tartaric acid has been introduced [1, 3, 4, 6]. The aim of this experiment was to compare the above technologies in terms of corrosion resistance and to compare the operating costs of individual technologies.

3.1 Material

The most frequently used series of materials include the 2000 series and the 7000 series. For this reason, the 2024-T3 material (AlCu4Mg1) and the 7075-T6 material (AlZn6Mg2Cu) were chosen for the experiment.

3.2 Technological Process

For the sake of the experiment a fully automated anodizing line was used. The line uses two-stage rinsing for greater water savings and better rinsing. The test specimens were sealed in demineralized water. The mixing of the baths was ensured by means of compressed air from the central distribution system.

Chromic Acid. The technological procedure of anodic oxidation using chromic acid is recorded in Table 1. The voltage of 22 V was used during anodic oxidation. The chemical composition of individual baths is recorded in Tables 2, 3 and 4. The average thicknesses of the anodic oxidation layer for material 2024 and 7075 were 3 μm (Table 5) [3, 5].

Table 1. Technological procedure of anodic oxidation in chromic acid

	Time [min]	Temperature [°C]
Alkaline degreasing	12	55
2° Rinse	6	–
Deoxidation	8,5	–
2° Rinse	11	–
Anodic oxidation	60	35,1
2° Rinse	6	–
Sealing	25	90,5
Drying	12,5	40,1

Table 2. Chemical composition of chromic acid bath

Anodic oxidation in chromic acid	Concentration
Total chromium oxide	61,50 $g \cdot l^{-1}$
Chromium oxide free	43,20 $g \cdot l^{-1}$
Al	2,21 $g \cdot l^{-1}$
pH	0,58

Table 3. Chemical composition of the alkaline degreasing bath

Alkaline degreasing	Concentration
Turco 4215 NC-LT	54,70 $g \cdot l^{-1}$
pH	8,70

Sulfuric Acid. The technological procedure of anodic oxidation in sulfuric acid was similar to the previous section. The process parameters are recorded in Table 6. The anodic oxidation used a voltage of 16.5 V and sulfuric acid at a concentration of 191.5 $g \cdot l^{-1}$. The average value of the measured anodic oxidation layer for material 2024 was 11.9 μm and for material 7075 it was 31.5 μm (Table 7) [3, 5].

Table 4. Chemical composition of the deoxidation bath

Deoxidation	Concentration
Turco Smut NC	19,4%
Al	0,043 $g \cdot l^{-1}$

Table 5. Anodized layer thicknesses in chromic acid [μm]

Sample number/Material	1	2	3	4	5	Average value
2024	2,3	2,2	2,6	2,5	2,4	2,4
7075	2,7	3,3	3,0	2,8	3,2	3,0

Table 6. Technological procedure of anodic oxidation in sulfuric acid

	Time [min]	Temperature [°C]
Alkaline degreasing	12	55,1
2° Rinse	6	–
Deoxidation	8,5	–
2° Rinse	6	–
Anodic oxidation	48	20,2
2° Rinse	10	–
Sealing	25	90,8
Drying	15	40,1

Table 7. Anodized layer thicknesses in sulfuric acid [μm]

Sample number / Material	1	2	3	4	5	Average value
2024	11,4	12,1	12,4	12,3	11,5	11,9
7075	32,2	31,6	31,4	30,8	31,3	31,5

Boric Acid. The technological process was again similar to the sections above, the process parameters are recorded in Table 8. The anodic oxidation used a voltage of 15 V. The parameters of the anodic oxidation bath are reported in Table 9. The average value of the measured anodic oxidation layer for material 2024 was 2,0 μm and for material 7075 it was 3.5 μm (Table 10) [4, 5].

Table 8. Technological procedure of anodic oxidation in boric acid

	Time [min]	Temperature [°C]
Alkaline degreasing	12	54,8
2° Rinse	6	–
Deoxidation	7	–
2° Rinse	12	–
Anodic oxidation	22,5	28,1
2° Rinse	6	–
Sealing	25,2	90,5
Drying	120	Room temperature

Table 9. Chemical composition of boric acid bath

Anodic oxidation in boric acid	Concentration
Sulfuric acid	$43{,}00\ g \cdot l^{-1}$
Boric acid	$8{,}16\ g \cdot l^{-1}$
Al	$4{,}18\ g \cdot l^{-1}$

Table 10. Anodized layer thicknesses in boric acid [μm]

Sample number/Material	1	2	3	4	5	Average value
2024	1,8	1,6	2,4	2,3	1,9	2,0
7075	3,2	3,6	3,5	3,8	3,3	3,5

Tartaric Acid. The technology of anodic oxidation using tartaric acid was divided into two groups. In the case of samples for the corrosion test, the technological procedure was similar to previous processes conducted in this experiment. The samples that were used for the weight test and the adhesion test of the base coat (3 + 3 test samples for the material) were not sealed, after anodic oxidation there was a 2° rinsing and drying. The parameters of the technological process are written in Table 11. During the anodic oxidation, a voltage of 14 V was used, the chemical composition of the bath is recorded in Table 12. The resulting anodized layer thicknesses were measured and recorded in the Table 13 [3, 5, 6].

Table 11. Technological procedure of anodic oxidation in tartaric acid

	Time [min]	Temperature [°C]
Alkaline degreasing	15	55,1
2° Rinse	10	–
Deoxidation	8	–
2° Rinse	10	–
Anodic oxidation	25	37,1
2° Rinse	10	–
Sealing	25	90,1
Drying	15	40,6

Table 12. Chemical composition of tartaric acid bath

Anodic oxidation in tartaric acid	Concentration
Sulfuric acid	$41{,}20\ g \cdot l^{-1}$
Tartaric acid	$79{,}20\ g \cdot l^{-1}$
Al	$1{,}20\ g \cdot l^{-1}$

Table 13. Anodized layer thicknesses in tartaric acid (sealed) [μm]

Sample number/Material	1	2	3	4	5	Average value
2024	3,9	2,7	3,3	4,1	2,8	3,4
7075	4,8	5,3	4,7	5,6	4,5	5,0

The specification [7] prescribes the primer (a two-component epoxy paint in a 3:1 ratio) for the adhesion test. The paint was cured in an oven at 80 °C for 30 min. The measured values of the coating thickness were entered in Table 14.

4 Evaluation

Three types of tests were selected for sample evaluation. Specifically, these are the corrosion test, the weight test, and the adhesion test of the primer. The corrosion test

Table 14. Coating layer thickness [μm]

Sample number/Material	1	2	3	Average value
2024	17,5	19,6	19,1	18,7
7075	18,4	18,1	19,3	18,6

was performed in a corrosion chamber (salt spray) according to ASTM B 117-11. The samples were inclined at an angle of 6° and a 5% sodium chloride solution was sprayed on them. According to MIL-A-8625, it was then stipulated that the test specimens must meet a minimum limit of 336 h in salt spray. The experiment was terminated when the first corrosion point was identified [8, 9].

The adhesion test of the primer was performed according to the standard EN ISO 2409 - Paints: Grid test. The distance between the individual sections of the grid is determined in the standard according to the thickness of the coating and the type of substrate. There are 6 classification levels from 0 to 5, where 0 indicates that the edges of the sections are completely smooth, and no grid square is damaged. Grade 5 is then evaluated as a grade that cannot be classified even as grade 4 (the damaged area is greater than 35% but less than 65%) [10, 11].

The weight test was performed according to ASTM B 137-95. Test specimens must be weighed after anodizing before staining and sealing. After weighing the samples, it is necessary to remove the coating from the test sample, then wash, dry and weigh them. A solution of phosphoric acid and chromic acid is used to remove the layer. The test pieces are immersed in the solution for at least 5 min (but not more than 6 min) at 212 ± 10 °C. It is necessary to observe the composition of the solution according to the following parameters [12]:

- Phosphoric acid, 85% – 35 ml
- Chromic acid (CrO3) – 20 g
- Water – remaining volume up to 1 l (1000 ml)

This procedure is repeated until the coating is completely removed (sample weight is constant). The resulting bulk density of the coating is obtained by subtracting the weight of the sample before and after removing the coating in milligrams. This result is divided by the surface area in square feet [12].

4.1 Corrosion Test

Chromium acid anodic oxidation met the minimum required limit of 336 h according to MIL-A-8625 specifications for materials 2024 and 7075. The identification of the first corrosion point occurred at 1440 h for both materials.

Anodic oxidation with sulfuric acid also met the required limit of 336 h in salt spray for both materials. For material 7075, the exposure ended after 629 h when the first corrosion point was observed, for material 2024, the test was interrupted after 4071 h, when the first corrosion point was still not observed.

Likewise, the anodic oxidation in boric acid and tartaric acid met the minimum limit of 336 h in salt spray according to specification MIL-A-8625. The exposure was terminated after 1661 h (material 2024) and after 1181 h (material 7075) for AO in boric acid and after 1008 h (both materials) for AO in tartaric acid.

Table 15 records all salt spray exposure termination values after the identification of the first corrosion point. The results of the corrosion test can be seen in Fig. 1, 2, 3 and 4.

Table 15. Comparison of AO technologies - identification of the first corrosion point (* after 4071 h there was no identification of corrosion)

Technology	2024	7075	MIL-A-8625
AO in chromic acid	1440 h	1440 h	OK
AO in sulfuric acid	4071 h (*)	629 h	OK
AO in boric acid	1661 h	1181 h	OK
AO in tartaric acid	1008 h	1008 h	OK

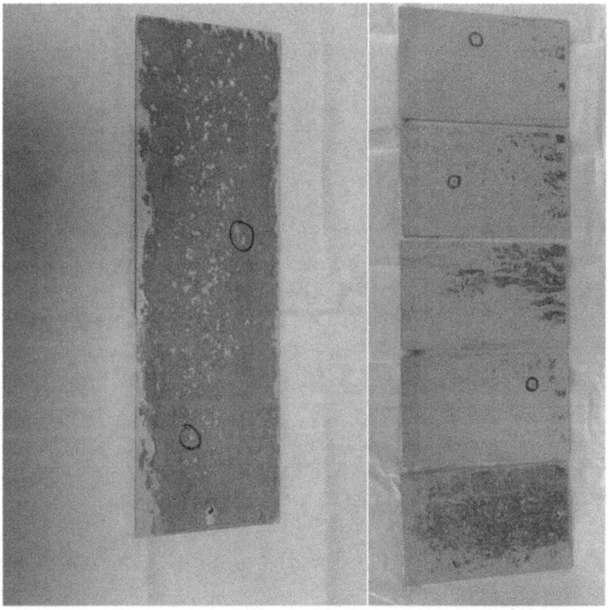

Fig. 1. Anodic oxidation using chromic acid after the corrosion test. From left a) material 2024 b) material 7075

Fig. 2. Anodic oxidation using sulfuric acid after the corrosion test. From left a) material 2024 b) material 7075

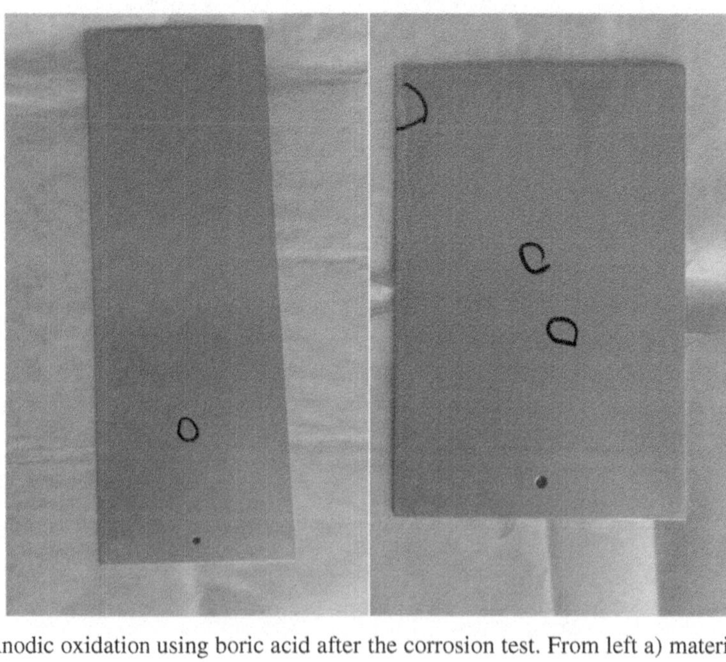

Fig. 3. Anodic oxidation using boric acid after the corrosion test. From left a) material 2024, b) material 7075.

Fig. 4. Anodic oxidation using tartaric acid after the corrosion test. From left a) material 2024, b) material 7075

4.2 Weight Test

The weight test was performed according to the ASTM B 137 specification, and it was also necessary for the samples to meet the requirements of the AIPI 02-01-002 specification, namely a minimum weight of 22 $mg \cdot dm^{-2}$ [7]. Both materials met the mentioned specification. Tables 16 (material 2024) and 17 (material 7075) record the initial and final sample weights and the weight calculation according to Eq. 1 (Table 17)

$$Weight = \frac{Initial\ weight - Final\ weight}{Sample\ area} \cdot 1000 \quad (1)$$

Table 16. Weight test results for mat. 2024

Sample	Initial weight[g]	Final weight[g]	Sample area[dm^2]	Weight[$mg \cdot dm^{-2}$]
1	25,80,345	25,73,876	1,2044	53,7092
2	25,87,737	25,81,455	1,2060	52,0895
3	25,83,648	25,77,312	1,2060	52,5373
Average weight				52,78

Table 17. Weight test results for mat. 7075

Sample	Initial weight[g]	Final weight[g]	Sample area[dm^2]	Weight[$mg \cdot dm^{-2}$]
1	25,94,999	25,85,938	1,2073	75,0548
2	25,70,001	25,60,823	1,2054	76,1415
3	25,82,754	25,73,767	1,2057	74,5381
Average weight				75,24

4.3 Base Coat Adhesion Test

The adhesion test of the base coat was carried out according to the standard EN ISO 2409 - Paints: Grid test. Both materials were classified as 0 "The edges of the cuts are completely smooth; no grid square is damaged" (Fig. 5).

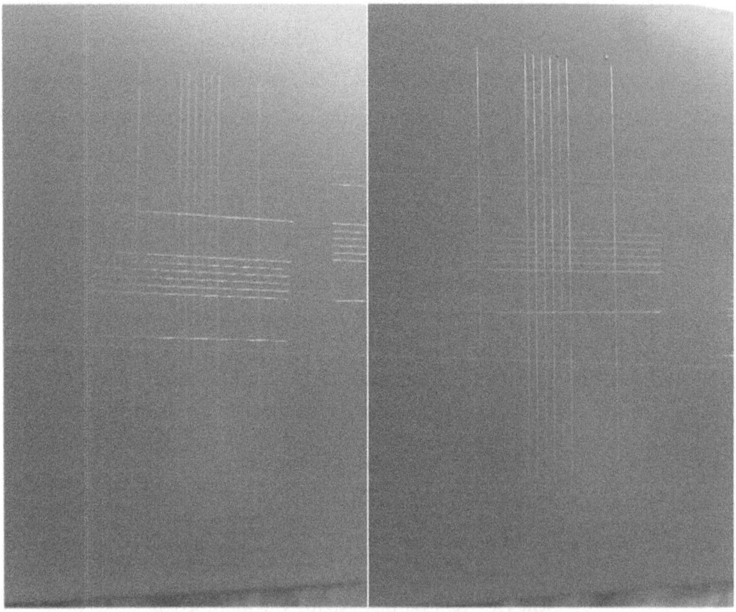

Fig. 5. Grid test: From left a) material 2024 b) material 7075

5 Economic Evaluation

The main idea is to find a suitable replacement for chromic acid, which would exhibit similar corrosion resistance while being environmentally friendly, and economically favorable. Therefore, this chapter deals with the economic comparison of individual baths.

The above-mentioned technologies have similar surface pretreatments and for this reason the economic evaluation was focused only on individual baths of anodic oxidation. The price is calculated for the volume of the bath 11,800 dm^3 for the given concentration specified in each of the sections, which pertains the technological process of specific acids. The order of the baths was recorded in Table 18.

Table 18. Economic evaluation – acid bath ranking [Kč]

Rank	Bath	Cost [Kč]
1	Boric acid	18 431
2	Sulfuric acid	32 214
3	Chromic acid	80 535
4	Tartaric acid	2634 654

6 Conclusion

This paper compared the corrosion resistance of four anodic oxidation baths. Chromic, sulfuric, boric, and tartaric acids were used. For tartaric acid, a weight test and an adhesion test using a grid test were also used. The anodic oxidation layer was formed on two types of material, namely 2024 and 7075.

In the corrosion resistance test, the samples were placed in a salt spray and left until the first corrosion point. A minimum limit of 336 h in salt spray was determined according to the MIL-A-8625 specification, which was met by test samples of all technologies and materials. The AO technology in sulfuric acid with material 2024 lasted the longest in the corrosion chamber. This combination still did not show any corrosion attack after 4071 h. AO in sulfuric acid shows the best results in corrosion resistance and is an ecological technology. Economically, this technology is one of the most demanding and there are design limitations due to the thicker anodic layer. Furthermore, this technology is not suitable to be used in the aerospace industry for stressed parts due to increased material fatigue. When comparing newer technologies (boric acid and tartaric acid) with those used for years (chromic acid and sulfuric acid). It can be seen from Tables 15 and 18 that tartaric acid has proven to be the least economically advantageous technology with the worst corrosion resistance. However, the advantage of this technology is that the fatigue of the material does not increase, and the bath does not contain hexavalent chromium and thus meets environmental demands. Boric acid has proven to be the most economically advantageous at a very low cost and meets the requirements of corrosion resistance (up to 1661 h in salt spray for material 2024). This technology has a very small layer thickness and therefore there is no design limitation. Boric acid baths also meet the requirement for the absence of hexavalent chromium.

Furthermore, a weight test was performed on AO samples in tartaric acid. For material 2024, a layer weight of 52.78 $mg \cdot dm^{-2}$ was measured. This value exceeds more than

2 times the required limit (22 $mg \cdot dm^{-2}$) determined by the AIPI specification 02-01-002. The weight of 75.24 $mg \cdot dm^{-2}$ was measured for material 7075. Here, too, the requirement of the specification was met, when the weight exceeds the required limit more than 3 times.

As the last experiment on AO samples in tartaric acid, the adhesion test of the base coat was performed using a grid test according to the EN ISO 2409 standard. The paint was cured in an oven at 80 °C for 30 min. The samples were classified as "0" and therefore it can be said that the anodizing layer with tartaric acid shows excellent surface adhesion for subsequent application of paints.

In conclusion, it is necessary to state that in terms of corrosion resistance, sulfuric acid appears to be the most advantageous. However, this acid is not used in the aerospace industry mainly due to design limitations imposed by the thicker layer and further increase in material fatigue. Boric acid meets the requirements for a very thin layer, low economic cost and good corrosion resistance. However, its worldwide use is being replaced by more environmentally friendly and safe technology anodic oxidation in tartaric acid. In this experiment, this acid proved to be the least economically advantageous technology with the worst corrosion resistance (it met the requirements of MIL-A-8625), but it does not increase material fatigue and it is a natural substance that is also used in the food industry. Therefore, it is a safe substance that is easily disposed of. Therefore, further testing of this acid is required to determine the usability of this substance in the aerospace industry as a possible replacement of chromic acid.

Acknowledgements. Research and experimental verification were carried out within the solution of the project CTU SGS22/OHK2/3T/12 - Influence of surface treatments on the quality of production technologies.

References

1. Juhl, A.D.: Overview of anodizing in the aerospace industry. Metal Finish. **108**(2), 20–21 (2010)
2. ECHA: European chemicals agency. https://echa.europa.eu/
3. González-Rovira, L., González-Souto, L., Astola, P., Bravo-Benítez, P., Botana, F.J.: Assessment of the corrosion resistance of self-ordered anodic aluminum oxide (AAO) obtained in tartaric-sulfuric acid (TSA). Surf. Coat. Technol. **399**, 31–39 (2020)
4. Zhang, J., et al.: The bonding strength and corrosion resistance of aluminum alloy by anodizing treatment in a phosphoric acid modified boric acid/sulfuric acid bath. Surf. Coat. Technol. **202**(14), 3149–3156 (2008)
5. Wasserbauer, J.: Anodic oxidation of aluminum alloys in the aerospace industry. CTU, Prague (2016)
6. Wang, R., Wang, L., Chunying, H.E., et al.: Studies on the sealing processes of corrosion resistant coatings formed on 2024 aluminium alloy with tartaric-sulfuric anodizing. Surf. Coat. Technol. **360**(9), 369–375 (2019)
7. AIPI 02-01-002: Sulfuric anodizing for aluminum
8. ASTM B117-19: Standard Practice for Operating Salt Spray (Fog) Apparatus. ASTM International, West Conshohocken, PA (2019)
9. MIL-A-8625. Military Specification: Anodic Coatings for Aluminum and Aluminum Alloys (1993)

10. EN ISO 2409. Paints - Grid test (2021)
11. Zoubek, M., Kudlacek, J., Chabera, P., Abramov, A.: Complex control method of degreasing process. In: Hamrol, A., Ciszak, O., Legutko, S., Jurczyk, M. (eds.) Advances in Manufacturing, pp. 575–585. Springer, Cham (2018). https://doi.org/10.1007/978-3-319-68619-6_55
12. ASTM B137-95: Standard Test Method for Measurement of Coating Mass Per Unit Area on Anodically Coated Aluminum, ASTM International, West Conshohocken, PA (2021)

Analysis of Emission Solid Particles from the 3D Printing Process

Anna Karwasz(✉) and Filip Osiński

Poznan University of Technology, Poznan, Poland
{anna.karwasz,fillip.osinski}@put.poznan.pl

Abstract. During additive manufacturing, particulate matter and volatile organic compounds are released. These pollutants contribute to irritation of the throat mucosa, cause cardiovascular disease and even stroke. The article determines the quantitative and mass share of particulate matter emissions during 3D printing from black ABS filament. The tests were carried out using an analyzer from the PEMS group. The obtained dust emission results were presented for both the entire processes and their most important parts. It was shown at which stage of additive manufacturing the emission is the highest. The greatest emission of pollutants occurred during printing, when samples were taken directly from the machine chamber. The PM mass emission intensity during printing reaches its highest values in the raft printing process. This is due to the large surface area of the material heated by the printing table during this operation, and thus the longer plasticization time of the ABS material itself. At the same time, the highest quantitative particle emission values occur in said raft printing step and during the heating of the nozzle and in the first phase of correct printing. It follows that precisely in these two periods there is a significant emission of small diameter (mass) particles. In rooms where we have many printers at our disposal, which print products simultaneously, special care should be taken. These rooms should be ventilated frequently or air purifying filters installed.

Keywords: Air pollution · Particulate matter · 3D printing · Human health

1 Introduction

Additive manufacturing is a field that is developing very quickly and dynamically. The rapid development of additive technologies has contributed to the decline in printer prices and increased their availability. Currently, 3D printers are common not only in offices or schools but also at home [1–3]. Due to short lead times and low costs, additive manufacturing is used in many areas of life. Just a few hours is enough to print the finished model with its final processing. 3D prints are used in medicine when diagnosing cancer of various internal organs, its location, shape or size [2–7]. 3D printing is used in unit or small lot production. It is an alternative to injection molding, which is characterized by a high cost of single or small-lot production, e.g. in the foundry industry [2, 8]. We also see the continuous development of additive manufacturing techniques by improving

B. Gapiński et al. (Eds.): MANUFACTURING 2022, LNME, pp. 216–226, 2022.
https://doi.org/10.1007/978-3-031-00805-4_18

print parameters using finite elements, genetic algorithms and artificial neural networks [9, 10].

The continuous development of technology causes, in addition to faster production of better products, a negative impact on the natural environment The effects of this negative impact on the natural environment and human health are systematically analyzed. During additive production, volatile organic compounds (VOCs) are released, which can irritate the mucosa of the respiratory tract, cause asthma and cardiovascular disease [11]. A significant threat to humans is the emission of solid particles (PM). The concept of solid particles is associated with air pollution that appears during combustion in domestic and industrial heating boilers and with cars equipped with internal combustion engines [12]. Due to air pollution, approximately 20,000 people die in Poland every year [13]. PM10 and PM2.5 particulate matter is the air pollutant that causes the greatest harm to human health. Some of its particles are so small (one-thirty to one-fifth the diameter of a human hair) that they not only penetrate deep into our lungs, but also enter our bloodstream, just like oxygen [14]. According to the report of the World Health Organization (WHO), over 90% of the urban population of the world breathes air that exceeds WHO's guideline levels for outdoor air pollution, and according to data from 2019, lung diseases are the 3rd cause of the largest number of deaths in the world [15].

The harmfulness of dust contained in the atmospheric air to human health depends on the physicochemical structure of the dust particles. The respiratory system is the main route of dust penetration into the human body. We can divide it into three functional areas, which differ in the time of dust retention in places of its deposition, pathological reaction to dust and the ways of its elimination. Particles with a diameter greater than 10 μm deposit in the upper parts of the respiratory tract, while PM2.5 particles reach the deeper parts of the lungs, where they are accumulated (see Fig. 1). The entire system responsible for delivering oxygen to the bloodstream has been divided into sections where solid particles can accumulate. Small particles pose the greatest threat to human health. Children and the elderly, as well as people with lung and heart diseases, are particularly exposed to the negative effects of solid particles in the atmosphere. The dust most dangerous to our health includes heavy metal particles (arsenic, lead, nickel, mercury, cadmium). Particles composed of heavy cyclic hydrocarbons (carcinogenic compounds) are also very toxic.

Small desktop-class 3D printers that can be used at homes have grown in popularity in recent years, but particles emitted from such devices can adversely affect indoor air quality and potentially harm health [17]. In the course of many years of research, various researchers collected particles emitted from 3D printers and conducted several tests to assess their effect on the respiratory tract [18]. Studies have shown a toxic reaction to particles from various types of filaments used in 3D printers. Researchers took a closer look at the chemical composition of the particles and their potential toxicity. During printing filament is heated, what causes numerous physical and chemical changes. As a result of thermal depolymerization and oxidation of the substances used, it releases, among others volatile organic compounds that are emitted into the air near the printer and printed object.

The aim of the study was to determine the quantitative and mass share of particulate matter emissions during 3D printing from black ABS filament. The tests were carried

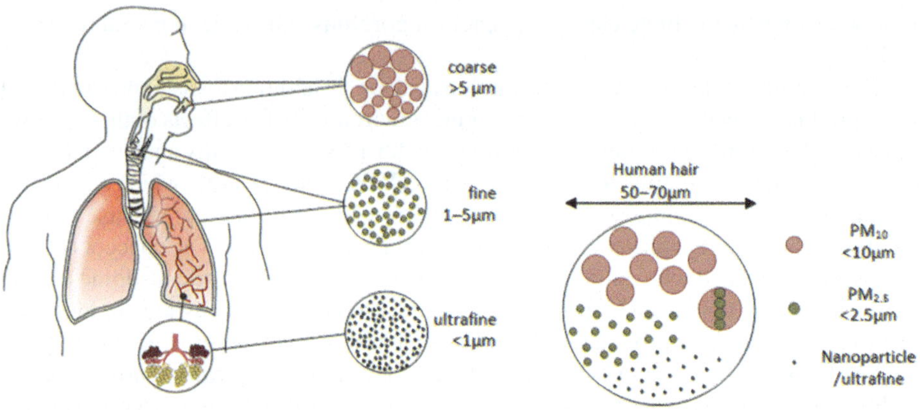

Fig. 1. Parts of the human respiratory system, particularly exposed to the effects of solid particles, depending on their dimensions [16].

out using an analyzer from the PEMS group. The research was to show at which stage of additive manufacturing the emission is the highest.

2 Materials and Methods

The first research object was the semiprofessional desktop-class printer Zortrax M 200 with a closed chamber, without a filter for cleaning the pollution. This printer is equipped with one extruder and a heated bed. It took an average of 60 min to print each test item. The research was done under standard conditions for ABS filament, in accordance with the manufacturer's recommendations. Filament used for tests was in black color. The most important parameters for printing from this filament are: nozzle temperature (275 °C), print bed temperature (80 °C), finished product mass (10 g), weight with raft (16 g) print time (68 min). No additional adhesive substances were used during printing. Filament moisture at the time of printing was below 0.2% and was in line with the recommendations for material moisture for ABS plastic processing. To standardize the results of the measurements, each test was carried out three times, each time cooling the print bed and hot end to the ambient temperature (cold start of the process).

The measuring station (see Fig. 2) consisted of a 3d printer that printed a given object (see Fig. 3). A mass spectrometer was connected to the printer (see Fig. 4), which examined the size (diameter) of the emitted pollutants and their mass during printing. A laptop was connected to the mass spectrometer, collecting and archiving measurement data.

Data was collected throughout the duration of additive manufacturing process. First, the measuring device was calibrated in an empty, ventilated room, with closed windows and doors, and then a printer and a laptop collecting data were connected. On the laptop screen, it was possible to observe the constantly changing values of the quantity and size of particles released during the printer's operation.

Fig. 2. Measurement site.

Fig. 3. Shape of the printed test product, finished product (8 cm × 5,3 cm × 0,8 cm).

There are many methods of measuring particulate emissions. Among others: a gravimetric method based on the analysis of particles collected by filtering using the differential weighing method.

Reference methods: manual (consists in gravimetric determination of the mass of dust accumulated on the filter as a result of passing through an unknown volume of air) and automatic (gravimetric method based on the sedimentation weight principle known mainly from TEOM type analyzers; radiation method based on radiation absorption β; optical methods laser diffraction, using the Fraunhofer theory and/or the MIE theory, and based on measurements in reflected light and transmitted light).

There is gas chronomatography, liquid chronomatography, mass spectrometer, pyrolysis, optical-electronic analyzer IPS P.

The measurement of particulate emissions during printing was performed using a mass spectrometer TSI Engine Exhaust Particulate Sizer 3090 (see Fig. 4) [19–24]. The devoice measures the concentration and dimensional distribution of solid particles from

5.6 to 560 nm with a time resolution of 10 Hz. The device can carry out measurements in dynamic and transient tests. Recorded data is constantly recalculated, among others on a linear and logarithmic scale. The concentration units are normalized (dN/dlogDp) for easy comparison with data obtained when measuring with other equipment [19].

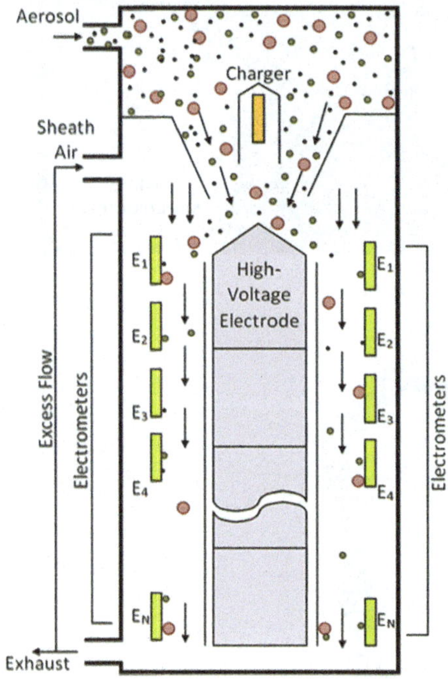

Parameter	Value
Size of measured particles	5,6 ÷ 560 nm
Number of measurement channels	16 channels for a decade (32 total)
Number of electrode channels	22
Resolution	10 channel sizes/second
Gas sample flow	10 dm3/min
Compressed air flow	40 dm3/min
Input sample temperature	10 ÷ 52°C
Device operating temperature	0 ÷ 40°C
Weight of entire device	32 kg

Fig. 4. Technical data of the mass spectrometer and mass spectrometer TSI EEPS 3090 for measuring the diametral distribution of solid particles [25].

The instrument measures the 32 channels of different size and concentration of particles, through the use of sensitive electrodes (see Fig. 4). The sensitivity of the spectrometer provides measurements from 200 pcs/cm^3, which corresponds to approximately <1 μg/m^3. Before the measurement, the exhaust sample is conditioned - at the measurement point by means of accessories or in a CVS tunnel (Constant Volume Sample). The tested gas volume flows through a filter, which captures particles larger than 1 μm, ie with a diameter outside the measurement range. Then the sample is led to the electrode chamber, where the solid particles receive the electric charge. This is where the ions are produced. As a result of the particles being charged, their dimensional classification can be carried out. The charging electrode is located on the right side of the device. From this point, the test particles pass into the annular gap, which is located between two coaxial cylinders. A stream of ambient air is supplied to the vicinity of the slot. The cylinder consists of a stack of highly sensitive electrodes, insulated from each other, arranged in a ring. They are grounded at one end, and connected to sensitive amplifiers at the other. Such a structure makes it possible to generate an electric field between the cylinders. The ambient air along with positively charged solid particles contained in it is directed to the gap between the rings. The sensitive electrodes in this space receive signals from charged particles.

The instrument draws a sample of the exhaust flow into the inlet continuously. Particles are positively charged to a predictable level using a corona charger. Charged particles are then introduced to the measurement region near the center of a high-voltage electrode column and transported left size the column surrounded by HEPA-filtered sheath air. A positive voltage is applied to the electrode and creates an electric field that repels the positively charged particles outward according to their electrical mobility. Charged particles strike the respective electrometers and transfer their charge. A particle with higher electrical mobility strikes an electrometer near the right side; whereas, a particle with lower electrical mobility strikes an electrometer left side in the stack. This multiple detector arrangement using highly sensitive electrometers allows for simultaneous concentration measurements of multiple particle sizes [25].

3 Results

Before the actual test of pollutant emission from 3D printers, the value of the measurement background was determined, in which the products were printed. The examination was carried out with both closed and open windows. Then, the equipment was set to zero in the background of the environment, therefore the final value of particulate matter emission did not have to be additionally corrected by its value. Then, measurements were made during 3D printing with direct sampling from the machine and at a distance from it (about 2 m). The supplement was the measurement of environmental pollution after the 3D printing process was complete. The room where the printing process was carried out has a volume of 208 m^3.

The biggest emission of pollutants occurred during printing when samples were taken directly from the machine chamber. The average emission value of PM was 113.20 $\mu g/cm^3$, while average numerical value of emission was 1.04E + 06 #/cm^3. At the end of the printing process, air in the printing room contained 3.39E + 04 #/cm^3 solid particles. It is nearly four times the concentration that was recorded in the environment before printing. That is why it is so important that the rooms where the additive manufacturing is proceeded are well ventilated or the 3D printers are equipped with local exhausts or air purifying / filtering devices.

The PM emission intensity in terms of mass and number for the individual stages of the printing process is shown in Fig. 5 and Fig. 6. The presented relations, it follows that the intensity of the mass emission of PM during printing reaches the highest values in the process of raft printing (for 7.2E + 02 $\mu g/s$). It may be related to the large surface of the material heated by the print table during this operation, and thus the longer plasticization time of the ABS material itself. At the same time, the highest values of quantitative particle emissions occur in the mentioned stage (up to 4.1E + 06#/s) and during nozzle heating (up to 3.2E + 06 #/s) and in the first phase of proper printing (up to 3.7E + 06 #/s). It follows that it is in these two periods, a significant emission of small diameter (mass) particles occurs.

Figure 7 and Fig. 8 shows the total particle number emission. In the printing process, the largest number occurred in the diameter range of 9.31–10.08 nm, 2.2E + 08 # and

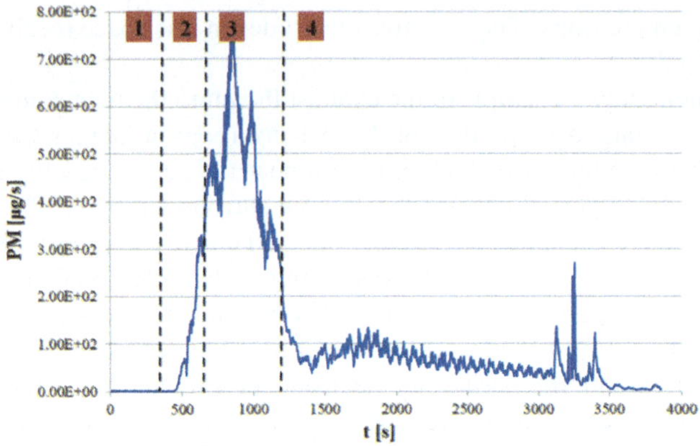

Fig. 5. PM 3D printer: 1-bed heating, 2-nozzle heating, 3- raft printing, 4-proper print.

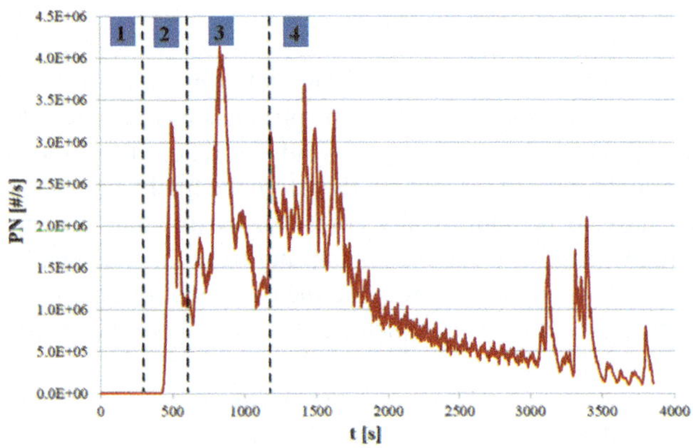

Fig. 6. PN 3D printer: 1-bed heating, 2-nozzle heating, 3- raft printing, 4-proper print.

2.1E + 08 #, respectively. At the same time, the obtained distribution was close to normal in the range of 16.5–124.1 nm.

4 Discussion

The conducted research is important from the point of view of individual users of desktop class 3D printers. The number of such devices is increasing each year in the use of both industrial and individual (home) users. This technology is becoming more and more popular every year and further growth in the development of this market is anticipated. It is estimated that in the years 2014–2017, the market value increased nearly 3 times from $ 3.07 billion to $ 8.68 billion [26].

Emission from the 3D printing process is also important due to the fact that due to the requirements of the process itself, it must take place in closed rooms where sudden air movements may not occur. This can cause an accumulation of harmful compounds. In the presented example, for a room with a volume of 208 m^3, the quantitative increase

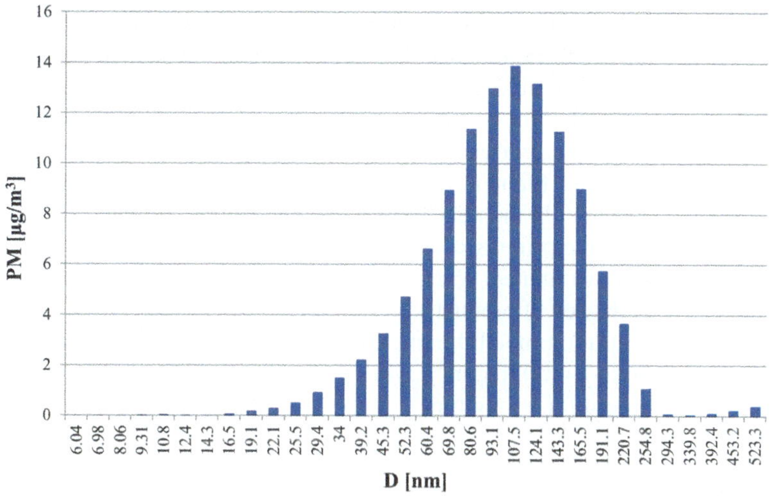

Fig. 7. Total emission of solid particles for the printing process.

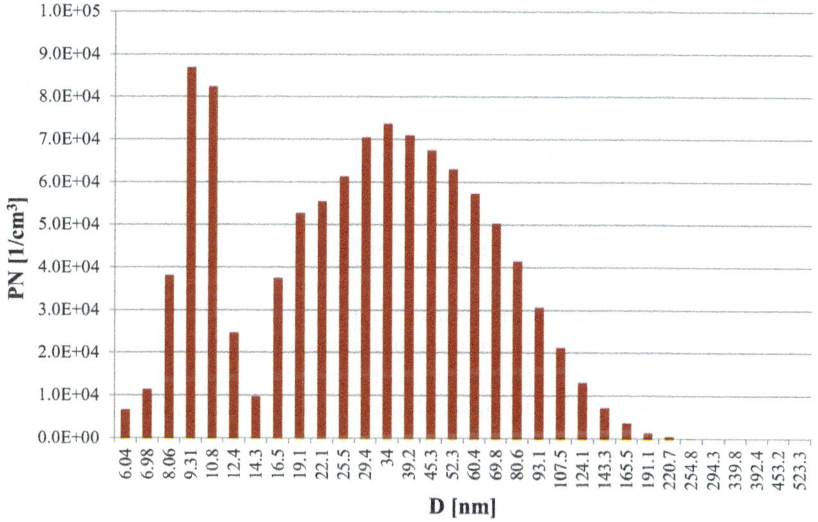

Fig. 8. Total emission of solid particles for the printing process.

in the concentration of solids after printing from a single device was increased 3 times. In the case of 3D printing, a relatively small increase in the mass of pollutants in the room is also important - it is mainly caused by the relatively small diameters of the particles emitted during the process. The vast majority of them were below 100 nm. As a result, these particles enter the alveoli without any major obstacles, and then into the bloodstream of the exposed person.

The conducted research contributes to wider research, namely a laboratory or a farm where there are more printing machines, for example 10–20 pieces. Especially that these pollutants can accumulate and remain in the room for a long time. The particles emitted during the process are small enough to observe Brownian motion in their behavior, which significantly reduces the possibility of their sedimentation over time. There are also no literature sources confirming the possibility of the accumulation of suspended particles, apart from organic solvents. For example, when printing with black ABS, after about 10 min, when the bed and filament are warmed up, there is a very large increase in the concentration of solids. In the next 10 min, the concentration reaches 720 [$\mu g/m^3$], Fig. 5. This is related to the printing of the raft on which the correct shape of the product is printed only after a few minutes. When printing the correct shape, the concentration of particulate matter decreases. If products are simultaneously printed on 10 or 20 printers in a room, the concentration of solid particles may increase from 720 [$\mu g/m^3$] to 7200 [$\mu g/m^3$] and even up to 14400 [$\mu g/m^3$]. After printing, the concentration of suspended dust remains in the room for a long time.

5 Conclusions

During the research, high emissions of nanoparticles were observed during the 3D printing process under normal operating conditions. A tendency to increase the concentration of solid particles in poorly ventilated rooms without additional air filters is a particular threat to health. Due to the fact that desktop 3D printers are increasingly used by home users, further research should be carried out, covering other materials used in the additive manufacturing process, and the impact of environmental conditions (including air humidity and the filament itself) on the risk posed by 3D printers should be verified for individual users.

Many authors who have been conducting research for many years confirm [18, 27–32, 34], that during additive manufacturing, volatile organic compounds are released into the environment.

The present studies confirm the measurements of other scientists [18, 30–34] that during 3D printing, volatile organic compounds with and small diameter particles (10–120 nm) are released. These particles can easily enter the body through the alveoli and further into the bloodstream. In addition, the color and type of used filament is of great importance, because of the coloring substances. In example the black filament is colored with soot.

Acknowledgments. The studies were realized with a support from statutory activity financed by Polish Ministry of Science and Higher Education (0613/SBAD/4710).

References

1. Żukowska, M., Górski, F., Hamrol, A.: Rating of polymers for low-cost rapid manufacturing of individualized anatomical models used in presurgical planning. In: Okada, H., Atluri, S.N. (eds.) ICCES 2019. MMS, vol. 75, pp. 633–644. Springer, Cham (2019). https://doi.org/10.1007/978-3-030-27053-7_54

2. Bernat, Ł, Kroma, A.: Application of 3D printing casting models for disamatch forming method. Arch. Foundry Eng. PAN **19**, 95–98 (2019). https://doi.org/10.24425/afe.2019.129637
3. Żukowska, M., Górski, F., Wichniarek, R., Kuczko, W.: Methodology of low cost rapid manufacturing of anatomical models with material imitation of soft tissues. Adv. Sci. Technol. Res. J. **13**, 120–128 (2019). https://doi.org/10.12913/22998624/113426
4. Jacek, B., et al.: 3D printed models in mandibular reconstruction with bony free flaps. J. Mater. Sci. - Mater. Med. **29**(3), 1–6 (2018). https://doi.org/10.1007/s10856-018-6029-5
5. Górski, F., Wichniarek, R., Kuczko, W., Żukowska, M., Lulkiewicz, M., Zawadzki, P.: Experimental Studies on 3D printing of automatically designed customized wrist-hand orthoses. Materials, 13, 4091–1–4091–19 (2020). https://doi.org/10.3390/ma13184091
6. Łukaszewski, K., Wichniarek, R., Górski, F.: Determination of the elasticity modulus of additively manufactured wrist hand orthoses. Materials, **13**, 4379–1–4379–18 (2020). https://doi.org/10.3390/ma13194379
7. Wierzbicka, N., Górski, F., Wichniarek, W., Kuczko, W.: Prototyping of individual ankle orthosis using additive manufacturing technologies. Adv. Sci. Technol. Res. J. **11**, 283–288 (2017). https://doi.org/10.12913/22998624/76070
8. Kuczko, W., Wichniarek, R., Górski, F., Buń, P., Zawadzki, P.: Application of additively manufactured polymer composite prototypes in foundry. Adv. Sci. Technol. Res. J. **9**, 20–27 (2015). https://doi.org/10.12913/22998624/2360
9. Górski, F., Wichniarek, R., Kuczko, W., Hamrol, A.: Selection of fused deposition modeling process parameters using finite element analysis and genetic algorithms. J. Multiple-Valued Logic Soft Comput. **32**, 293–311 (2019)
10. Rojek, I., Mikołajewski, D., Dostatni, E., Macko, M.: AI-optimized technological aspects of the material used in 3D printing processes for selected medical applications. Materials, **13**, 5437–1–5437–19 (2020). https://doi.org/10.3390/ma13235437
11. Karwasz, A., Osiński, F.: Literature review on emissions from additive manufacturing by FDM method and their impact on human health. Manag. Prod. Eng. Rev. **11**, 65–73 (2020). https://doi.org/10.24425/mper.2020.13493
12. Merkisz, J., Pielecha, J.: Particulate matter emissions from automotive sources. Poznan University of Technology Publishing House, Poznań (2014). (in Polish)
13. Adamkiewicz, Ł.: External health costs of air pollutant emissions from the household and municipal sector. Ministry of Entrepreneurship and Technology (2016). (in Polish)
14. Kocbach, A., Li, Y., Yttri, K.E., Cassee, F.R., Schwarze, P.E., Namork, E.: Physicochemical characterisation of combustion particles from vehicle exhaust and residential wood smoke. Particle Fibre Toxicol. **3** (2006). https://doi.org/10.1186/1743-8977-3-1
15. www.who.com. Accessed 03 Jan 2021
16. Pośniak, M.: Diesel engine exhaust - principles and methods of occupational exposure assessment. Medycyna Pracy, **54**, 389–393 (2003). (in Polish)
17. Every breath we take. Improving air quality in Europe, EEA Signals, Copenhagen (2013). https://www.eea.europa.eu/publications/eea-signals-2013. Accessed 23 Oct 2020
18. Zhang, Q., et al.: Chemical composition and toxicity of particles emitted from a consumer-level 3D printer using various materials. Environ. Sci. Technol. **53**, 12054–12061 (2019). https://doi.org/10.1021/acs.est.9b04168
19. Warguła, Ł., Kukla, M., Lijewski, P., Dobrzyński, M., Markiewicz, F.: Impact of compressed natural gas (CNG) Fuel systems in small engine wood chippers on exhaust emissions and fuel consumption. Energies **13** (2020)
20. Szymlet, N., Lijewski, P., Kurc, B.: Road tests of a two-wheeled vehicle with the use of various urban road infrastructure solutions. J. Ecol. Eng. **21**, 152–159 (2020). https://doi.org/10.12911/22998993/125503

21. Rymaniak, Ł, et al.: The role of real power output from farm tractor engines in determining their environmental performance in actual operating conditions. Comput. Electron. Agric. **173**, 105405 (2020). https://doi.org/10.1016/j.compag.2020.105405
22. Merkisz, J., Gallas, D., Siedlecki, M., Szymlet, N., Sokolnicka, B.: Exhaust emissions of an LPG powered vehicle in real operating conditions. In: E3S Web of Conferences, vol. 100. EDP Sciences (2019)
23. Lijewski, P., Fuc, P., Dobrzynski, M., Markiewicz, F.: Exhaust emissions from small engines in handheld devices. In: MATEC Web of Conferences, vol. 118 (2017). https://doi.org/10.1051/matecconf/201711800016
24. Merkisz, J., Lijewski, P., Fuc, P., Siedlecki, M., Weymann, S.: The use of the PEMS equipment for the assessment of farm fieldwork energy consumption. Appl. Eng. Agric. **31**, 875–879 (2015)
25. Pielecha, J.: Testing the emission of pollutants from internal combustion engines. Poznan University of Technology Publishing House, Poznań (2017)
26. Attaran, M.: The rise of 3-D printing: the advantages of additive manufacturing over traditional manufacturing, vol. 60, pp. 677–688. Business Horizons (2017). https://doi.org/10.1016/j.bushor.2017.05.011
27. Hall, S., Pengelly, I., Staff, J., Plant, N., Evans, G.: Measuring and controlling emissions from polymer filament desktop 3D printers, Research Report, Health and Safety Executive (2019). https://www.hse.gov.uk/research/rrhtm/rr1146.htm. Accessed 03 July 2021
28. Stefaniak, A.B., et al.: Characterization of chemical contaminants generated by a desktop fused deposition modeling 3-dimensional Printer. J. Occup. Environ. Hyg. **14**, 540–550 (2017). https://doi.org/10.1080/15459624.2017.1302589
29. Steinle, P.: Characterization of emissions from a desktop 3D printer and indoor air measurements in office settings. J. Occup. Environ. Hyg. **13**, 121–132 (2016). https://doi.org/10.1080/15459624.2015.1091957
30. Azimi, P., Zhao, D., Pouzet, C., Crain, N.E., Stephens, B.: Emissions of ultrafine particles and volatile organic compounds from commercially available desktop tree-dimensional printers with multiple filaments. Environ. Sci. Technol. **50**, 1260–1268 (2016). https://doi.org/10.1021/acs.est.5b04983
31. Yi, J., et al.: Emission of particulate matter from a desktop three-dimensional (3D) printer. J. Toxicol. Environ. Health A **79**, 453–465 (2016). https://doi.org/10.1080/15287394.2016.1166467
32. Potter, P.M., Al-Abed, S.R., Lay, D., Lomnicki, S.M.: VOC emissions and formation mechanisms from carbon nanotube composites during 3D printing. Environ. Sci. Technol. **53**, 4364–4370 (2019). https://doi.org/10.1021/acs.est.9b00765
33. Youn, Jong-Sang., Seo, Jeong-Won., Han, S., Jeon, Ki-Joon.: Characteristics of nanoparticle formation and hazardous air pollutants emitted by 3D printer operations: from emission to inhalation. RSC Adv. **9**(34), 19606–19612 (2019). https://doi.org/10.1039/C9RA03248G
34. Gu, J., Wensing, M., Uhde, E., Salthammer, T.: Characterization of particulate and gaseous pollutants emitted during operation of a desktop 3D printer. Environ. Int. **123**, 476–485 (2016). https://doi.org/10.1016/j.envint.2018.12.014

Computer Control Systems Used in Precision Agriculture

Mateusz Wiecha[1], Mariusz R. Rząsa[1(✉)], and Karolina Adamczyk[2]

[1] Opole University of Technology, 45-271 Opole, Poland
m.rzasa@po.opole.pl

[2] Schools Complex of the Agricultural Training Center named by Fr. Jan Dzierzon in Bogdańczowice, Bąków, Poland

Abstract. The paper presents a computer system for monitoring plant growth, developed for the needs of precision agriculture for small agricultural areas. The work contains a description of the monitoring system with a breakdown into the key elements of the process. An exemplary method of preparing orthophotomaps of the area was presented. The method of making maps that can be implemented on a PC computer has been described. The paper describes the most frequently used Vegetation Index. A test of determining the coefficients was carried out on an exemplary aerals with an area of 5.28 ha. Typical positioning systems for agricultural machines are discussed. The DGPS navigation method was used in the tests. Tests have confirmed that it can be used in precision agriculture with small aerals. The solution is optimal in terms of positioning accuracy and economics of small farms. The presented system was tested during one cycle of vegetation of winter barley sown with the no-plowing method. On this basis, the complexity of the system was assessed and its implementation was proposed. The proposed solution does not require complex computer systems. It has been designed so that it can be implemented on standard PC equipment cooperating with a short-range drone equipped with a standard RGB camera.

Keywords: Precision agriculture · Field mapping · Vegetation indicator · Geological indicator · Precision GPS

1 Introduction

Agricultural production has been an important branch of the economy for many years. The efficiency of agricultural production is an important element in this type of production. We now distinguish broadly between two types of farming: extensive and intensive [1].

Extensive farming - is a way of cultivation in which no artificial crop supports such as mineral fertilizers and pesticides are used, which however results in reduced yields. In this system, labor effort is high and costs are low. This type of farming is applied in less economically developed countries or where there is a large agricultural area. In the world it is very popular in the Middle East, Central America, United States and Africa

B. Gapiński et al. (Eds.): MANUFACTURING 2022, LNME, pp. 227–238, 2022.
https://doi.org/10.1007/978-3-031-00805-4_19

mainly in the tropics throughout. In those places, there is no need for very high yields. Large harvests are achieved by growing large areas of land.

A variety of extensive agriculture which is used all across the world is ecological agriculture. It is based on the stimulation of agricultural production through the use of natural and technologically unprocessed resources. In this way, a very high productivity is not achieved, however, the high biological quality of the agricultural products is more important.

In highly developed countries with a relatively small average area of farmland, intensive agriculture is popular. It is especially popular in Western European countries. It allows high yields to be obtained by intensive fertilization, high mechanization and automation of production. It is a very reasonable solution wherever there is a high population density.

Intensive agriculture includes precision farming, which uses information technology to match fertilizer and plant protection product doses to the potential demands of the crop to be grown. Over the past few years, precision agriculture has been implemented into production in many countries on an ever-increasing scale. On several occasions, attempts have been made to combine agriculture with other areas. As an example, the process of utilization of biological waste and soil fertilization can be combined. Currently, composting is becoming more and more popular as a method of utilizing biological waste [2] and the ST-OP technology on the management of plant residues [3]. The composting process is completed with hygienisation, which makes the final product non-toxic and it is possible to spread it on less fertile agricultural land for enrichment before sowing crops.

Precise fertilization lowers production costs and reduces the environmental impact of pesticides. However, it is linked to the analysis of soil conditions that affect plant growth and the tracking of crop growth. This approach requires the use of appropriate measurement techniques to monitor crop growth. Such monitoring provides the ability to isolate areas to which the appropriate amount of fertilizers should be selectively delivered. In this work, it is presented as a solution for computer monitoring of agricultural acreage and an example of its implementation on a sample agricultural area.

2 Precision Agriculture System

The base of the management system for the cultivation process is the precise monitoring of the cropland which involves the use of GPS tracking and the plotting of high accuracy orthophoto maps which provide a true picture of the cultivated area [4]. Figure 1 shows the next steps of monitoring and controlling the process. A process begins by mapping the acreage before sowing with precise GPS positioning. Appropriate precision is obtained by reference markers in the field. For such a marked area, a picture of it shall be made using drones or airplanes. Subsequently, the picture is being computer analyzed for creating a map of the area's fertility. On this basis, selective fertility treatments are carried out to make the area homogenous. Then, after seeding, the vegetation is monitored by drones. Vegetation maps give the base for computer analysis, upon which it is decided on the selective fertilization, the application of appropriate plant protection measures and irrigation. The analysis algorithm shall take into account the weather conditions and their forecast. The monitoring process is carried out throughout the growing season of the crop. Following the end of the vegetation period, the harvesting process is carried out with monitoring of the productivity. Harvest data provides valuable information about the effectiveness of the process, and based on this information, it is possible to make adjustments to the computer algorithms that will be used in future plant growth control processes.

Precision farming systems combine knowledge of soil fertility (soil productivity) with the process of fertilization, application of plant protection products and agrotechnical practices. This process is supported by new informatics and tele-informatics technologies using appropriately adapted machines. Such a process allows for effective use of production resources, improving the volume and quality of yields while minimizing environmental damage.

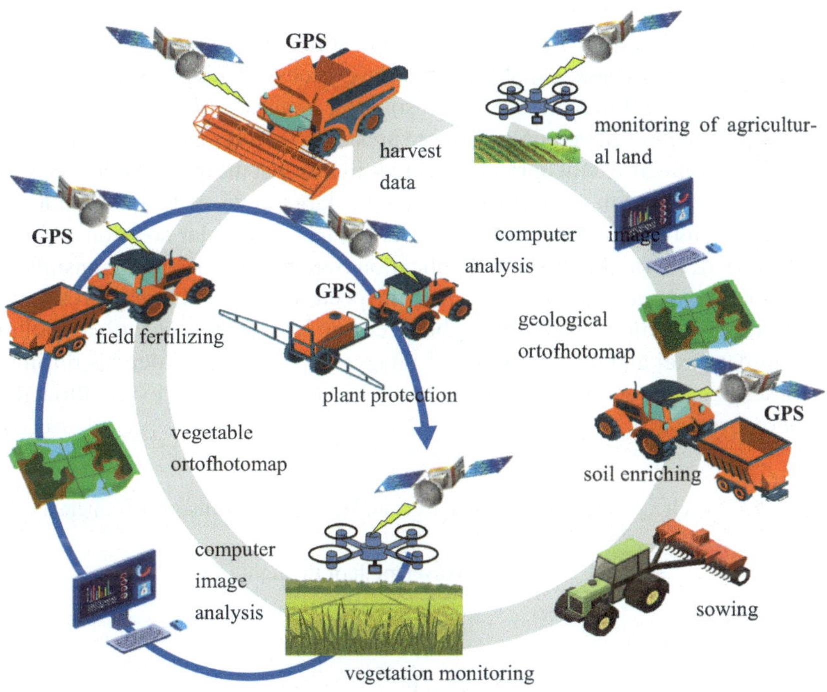

Fig. 1. Precision farming system duty cycle. (Graphics from vecteezy.com were used for the drawing).

Capturing information and documenting it forms the basis of predictive algorithms used in precision agriculture. Information on spatial and temporal variability of characteristics of plants in combination with weather conditions provides a knowledge base for predictive algorithms based on neural networks. Repeatedly, these data are being supplemented by information on the occurrence of plant macrophages that occur in a particular area, which allows for much more precise selection of plant protection products. The information are usually collected under the form of spatial-temporal maps.

As the key element of the management system in precision agriculture, is the monitoring process for the preparation of precise orthophoto maps of the terrain, associated with a process of image analysis that permits the identification of zones requiring intervention. Secondarily, an important element is a precision guidance system for agricultural machinery in areas requiring interventions. The following sections of this article will discuss the key elements of this system.

3 Technology of Orthophotomaps Production

Orthophotomaps are prepared on the basis of acreage photographs. These days it is possible to take pictures of an area of several hundred or even several thousand hectares with the use of satellite techniques. Such pictures are characterized by very high resolution and may be used to monitor the growth of plants. However, the access to these images is very limited. Mostly military organizations and some state institutions have access to these images. Farm owners usually do not have access to high-resolution satellite images. They are only allowed to use low-resolution images. This type of images can be used for assessment of lichen on very large areas, up to thousands of hectares.

Regarding the typical size of agricultural areas in Europe, the usage of this technique is marginal. For preparation of orthophotomaps for large areas of tens to hundreds of hectares a good solution is taking pictures from a plane. A popular solution is ATLAS system (Advanced Thermal and Land Application Sensor). The system is installed on a light airplane, the device is a 15-channel multispectral scanner that captures thermal radiation, infrared radiation and visible light spectrum. It takes images with a resolution of about 2 m per pixel and takes a series of images at one-second intervals during the flight. In order to create an orthophotomap of an area it is necessary to put altogether a series of photographs [5].

Table 1. Wavelengths recorded in images from Hyperspectral Imaging Cameras.

Color	Wavelength of the light [nm]
B - Blue	0.4826
G - Green	0.5613
R - Red	0.6546
NIR – Near Infrared	0.8646
SWIR 1	1.6090
SWIR 2	2.2010

The position of the aircraft, its orientation and the sensor orientation are all recorded at least once a second. The active calibration and record of a position mean enable accurate and reproducible measurement of the field plots while flying in a jet aircraft.

ATLAS is able to sense 15 multispectral radiation channels across the thermal – near in-frared – visible spectrums. The sensor also incorporates onboard, activates calibration sources for all bands. Atlas is capable of approximately 2.0 m resolution per pixel when flown in NASA's Lear jet and sees about a 30 degree swath width to each side of the air-craft.

For areas ranging from a few hectares to several tens of hectares, a good solution is to use a small camera with a broad-spectrum camera. It is a very practical solution because of the typical size of agricultural areas in Europe. Similarly as in the case of pictures taken from a plane, it is required to combine them into one whole. Typical resolutions are several to several dozen cm/pixel.

Hyperspectral Imaging Cameras are used to take pictures [6]. These cameras record images at different wavelengths of light from visible light to deep infrared (Table 1).

Camera images are usually presented in two color compositions (Fig. 2). An RGB composition is created when the three basic RGB channels are assigned the actual spectral colors according to Table 1. A characteristic feature of this type of painting is that the color of the water (e.g. a river) (Fig. 2a) takes on a dark green tone and is low contrast against the vegetation. To improve the contrast of elements in the image, a color conversion is performed in the unreal CIR composition. It consists in replacing values of RGB channels with values from other wavelengths. Color B is assigned to values corresponding with Green (0.5613), color G is assigned to values corresponding with Red (0.6546) and color R is assigned to NIR values (0.8646). For the example map in Fig. 2b. The river is colored black or dark blue, while the intense vegetation is colored red, which is quite a high contrast.

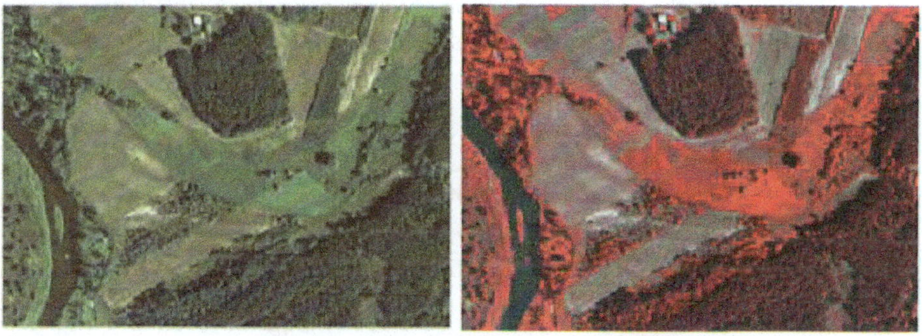

a) b)

Fig. 2. Sample picture from color composition [7] a) RGB b) CIR.

Besides the presentation of orthophotomaps an equally important process is assembling the sub-images coming from the camera placed on the drone or plane into one map. The particular images need to be scaled and their position needs to be adjusted. For this purpose, subsequent images are added to the orthophotomap and then adjusted. In order to match the positions of individual images, there is used the best-correlated pixels method [8]. It consists in searching for the best correlation both with the pixel in the Y_k image with the same coordinates as the pixel in the X_k image and with the pixels in its immediate neighbourhood (Fig. 3). The final result is the shift value for which the highest correlation coefficient is obtained.

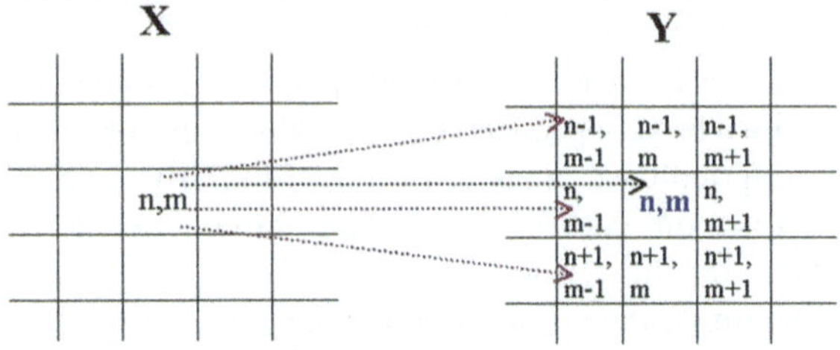

Fig. 3. The best-correlated pixels method.

The mathematical notation of this method with the restriction of the distance of neighboring pixels to pixels from the set B is expressed by the equation and the calculation of the largest correlation value R_X:

$$R_{X_{k[m,p]},Y_{k[m+l,p+j]}}[n] = \sum\nolimits_{l \in B} \sum\nolimits_{j \in B} \sum\nolimits_{n=0}^{N_k-1} x_k[m,p] \cdot y_{(k+n)}[m+l,p+j] \quad (1)$$

where:

N – number of pixels in the overlap area

x_k, y_k – pixel values for the offset k of coordinates [n,m]

Regardless of the correct image combination, there is a common area called overlap (Fig. 4). Because with the change in position of the plane or drone, the illumination of the photographed object changes, it is necessary to correct the colors appropriately.

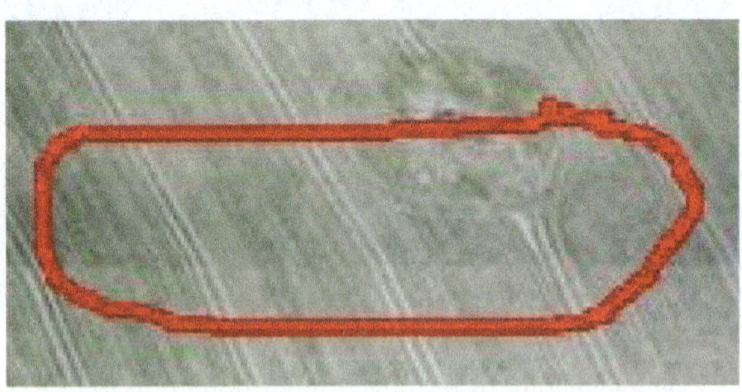

Fig. 4. The overlap area.

Color correction corrects the intensity of the RGB components in the add image. The correction algorithm consists of calculating the averaged correction coefficients for each RGB component, and then multiplying the values of the RGB components in all pixels of the Y-added image. The correction factors are calculated using the relationship:

$$k_B = \frac{1}{N} \sum\nolimits_{i=1}^{N} \frac{B_{Xi}}{B_{Yi}} \quad (2)$$

where:

B_{Xi}, B_{Yi} – values of the B component in i-th pixel

4 Precise Guidance System for Agricultural Machinery

To implement precise silvicultural treatments based on orthophotos, it is necessary to have a precession guidance system for agricultural machinery in the field. Since 2000, the signal from 24 NAVSTAR satellites has been widely shared. However, the U.S. Department of Defense disabled the SA (Selective Availability) module, whose task was intentional interference of signals causing a decrease in tracking accuracy from 3 to 12 m. Guidance is based on receiving signals from several NAVSTAR satellites with known coordinates and on their basis calculating the position of the agricultural machine (Fig. 5).

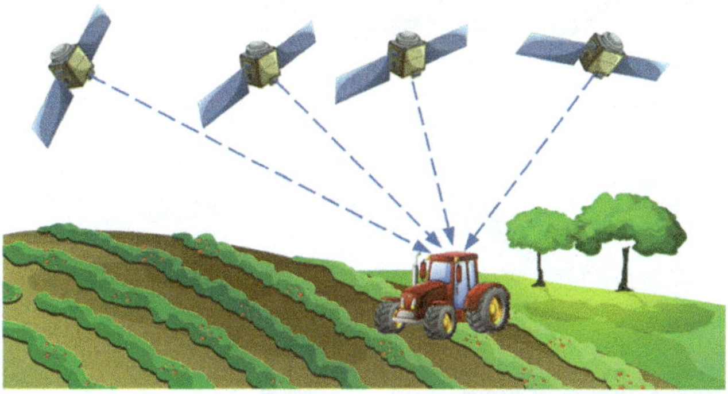

Fig. 5. GPS navigation (Graphics from vecteezy.com were used for the drawing)

This solution is unsuitable for precision agriculture where accuracy of less than 1m is required. To obtain an accuracy of less than 1m, correction of the GPS signal by external assistance systems called Differential GPS (DGPS) is required (Fig. 6).

DGPS systems send their own signals to the receiver, which contain differential corrections to correct the position calculation error [9]. In DGPS systems differential corrections are sent by land base (reference) stations with precisely known position (Fig. 6a) or by geostationary satellites (Fig. 6b). In the solution of Fig. 5a the range of correction signals is limited to a few or several kilometers, depending on the power of the base station transmitter. Much larger, because with global network coverage, communication with the ground station via geostationary satellites is possible. Currently, there are 5 differential satellite systems operating in the world, which correct GPS signals: WAAS – covering the territory of North America; EGNOS – covers Europe which can be supported additionally by the Russian system GLONASS and MSAS – involving Japan. In addition, there are paid global OmniSTAR and StarFire (from JohnDeere) systems which are available. Standard DGPS systems send differential corrections with some delay due to the need for data post-processing. Therefore, the positioning accuracy for one-channel DGPS systems (EGNOS, WAAS, OmniSTAR-VBS, StarFire-1) is

0.5–1.0 m. For two-channel systems (Omni-STAR-HP, StarFire-2), where the signal is emitted in two frequency bands with simultaneous use of ground reference stations, the positioning accuracy is 10–15 cm.

a)

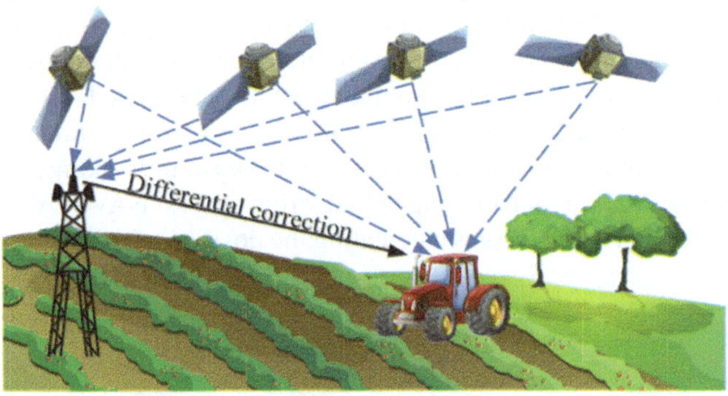

b)

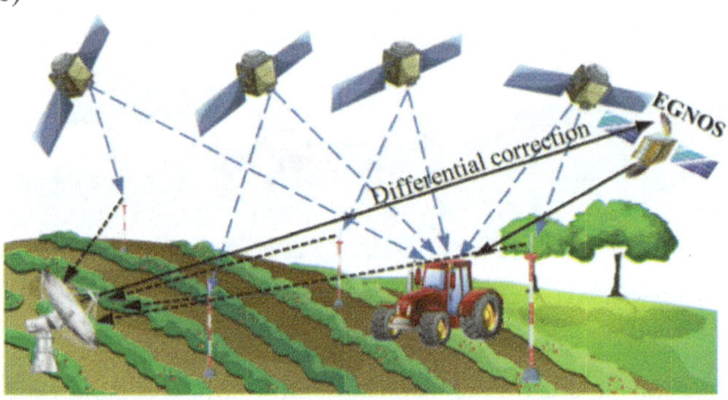

Fig. 6. DGPS navigation (Graphics from vecteezy.com were used for the drawing)

The most advanced tracking system is RTK-DGPS (Real Time Kinematic DGPS), where correction signals are received from a reference station in real time (Fig. 7). They allow for positioning accuracy of 1–3 cm. However, it requires setting up additional transmitting antennas in close proximity to agricultural machinery.

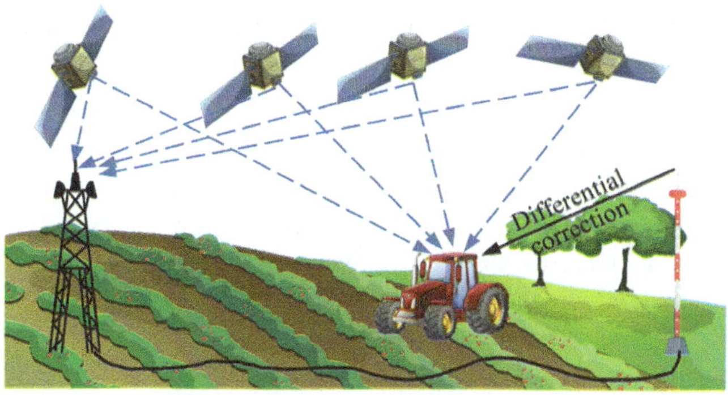

Fig. 7. RTK-DGPS navigation (Graphics from vecteezy.com were used for the drawing)

5 Plant Growth Monitoring

The vegetation index is a parameter that determines the quality of the growth of plants. Based on this index, a decision concerning the use of fertilizers or plant protection products can be made. It should be determined periodically during the vegetation process. If significant differences in the growth of plants are observed on a given area, it is necessary to go to selected places to take samples for laboratory tests. Based on the samples taken from the selected sites, the cause of the reduced growth is determined and a follow-up strategy is developed.

Defining the vegetation index involves measuring the difference between light absorbed and reflected by plant leaves. The ratio of light absorbed to light reflected from the plant surface is influenced by the presence of chlorophyll, water and cellular structures. Chlorophyll absorbs red light at about 0.67 nm and blue light at about 0.45 nm. The contained water in the leaves absorbs infrared light, with a wavelength of about 1.4 to 1.9 nm, and the cellular structures strongly reflect wavelengths in the range of 700 to 1100 nm. This provides the basis for developing a range of vegetation indices.

One of the most popular vegetation indices is the Normalized Difference Vegetation Index (NDVI), which measures the ratio of the difference and sum of the amount of reflected near-infrared and red light. The basis of this formula is the absorption of red light by chlorophyll and the low absorption of the near-infrared range by green leaves [10].

$$NDVI = \frac{NIR - RED}{NIR + RED} \tag{3}$$

To detect areas where plant germination does not occur, a good indicator is the SR (Simple Ratio). It is determined from the ratio between the amount of reflected near-infrared and red radiation. For green plant leaves this ratio can take on values of several tens, while for unplanted soil it is about zero [11].

$$SR = \frac{NIR}{RED} \tag{4}$$

Soil background makes it difficult to observe vegetation based on indicators. This is especially difficult when plant vegetation that partially covers the soil is observed. In this case, it is reasonable to use the Modified Soil Adjusted Vegetation Index (MSAVI). The disadvantage of this index is its sensitivity to changes in the atmosphere [12].

$$MSAVI = \frac{2 \cdot NIR + 1 - \sqrt{(2 \cdot NIR + 1)^2 - 8(NIR - RED)}}{2} \tag{5}$$

To reduce the impact of atmospheric conditions, it has developed the Visible Atmospherically Resistant Index (VARI). The advantage of this solution is analysis in the range of visible light, which enables to use RGB cameras to monitor the area [13].

$$VARI = \frac{GREEN - RED}{GREEN + RED - BLUE} \tag{6}$$

a) b)

Fig. 8. An orthophotomap of the study aerial taken in the spring after barley germination (a) RGB image (b) distribution of VARI index values

An exemplary analysis of barley vegetation on the tested area was made for an orthophoto taken in spring (Fig. 8a). From that, the VARI index was calculated and its values are presented on Fig. 8b.

As it was done for the IOR indicator after Image Processing, there are determined the areas which require detailed analysis and the application of appropriate measures to improve the growth in these areas (Fig. 9).

Fig. 9. Orthotopmap of test aerate before barley feed

6 Conclusions

The system presented in this publication can be used to monitor small agricultural areas. This solution is especially important due to the fact that nowadays precision farming for small agricultural areas is relatively unpopular among farmers. One of the reasons is the relatively high cost of commercial systems that are designed for large areas. The results shown in this report should be used to build low-cost systems for precision agriculture.

On the basis of the performed tests of making orthophotos from fragmentary photographs of the area, it can be concluded that with the use of typical tools for assembling images it is possible to make maps of this type on PC computers. The tests used The best-correlated pixels method to position fragmentary photos. This is the solution that is used in most panoramic photo editor programs. This is an example of the use of typical applications to make orthophotos. Similarly, typical graphics software is equipped with color correction algorithms. The conducted tests showed the possibility of using this type of algorithm to prepare orthophotos for the purposes of precision agriculture. Similarly, the use of the DGPS navigation method confirms its usefulness in this type of solution. To sum up, for small farms it is possible to create a low-cost solution based on short-range drones, PC-class computers and RGB cameras for monitoring plant growth in precision agriculture. Such solutions can be promoted in this area of agriculture.

In agricultural practice, the possibility of creating a discretized orthophotomap is very helpful. This type of map allows you to control the agricultural sprayer start-up system. Currently, the connection of information from the GPS system installed on agricultural machines with the control mechanism of the sprayer valve or starting the spreader is not a major problem. Retrofitting the system with tools for preparing orthophotomaps enables a fuller use of the possibilities of agricultural machines used in small farms.

References

1. Sharifi, P.: Sustainable agriculture: and introduction to extensive and intensive agriculture. J. Eng. Appl. Sci. **12**(10), 2747–2751 (2017). https://doi.org/10.36478/jeasci.2017.2747.2751
2. Anders, D., Rząsa, M.R.: The possibility of composting animal waste products. Environ. Prot. Eng. **33**(2), 7 (2007)
3. Jaskulska, I., Romaneckas, K., Jaskulski, D., Wojewódzki, P.: A strip-till one-pass system as a component of conservation agriculture. Agronomy **10**(12), 2015 (2020). https://doi.org/10.3390/agronomy10122015
4. Evcim, H.Ü., Değirmencioğlu, A., Özgünaltay, E.G., Aygün, İ: Advancements and transitions in technologies for sustainable agricultural production. Econ. Environ. Stud. **12**(4), 459–466 (2012)
5. Barwicki, J., Mazur, K., Wardal, W.J., Majchrzak, M., Borek, K.: Monitoring of typical field work in different soil conditions using remote sensing – a literature review and some concepts for the future. Agric. Eng. **19**(3, 155), 5–13 (2015)
6. Tahmasbian, I., Morgan, N.K., Bai, S.H., Dunlop, M.W., Moss, A.F.: Comparison of hyperspectral imaging and near-infrared spectroscopy to determine nitrogen and carbon concentrations in wheat. Remote Sensing **13**, 1128 (2021). https://doi.org/10.3390/rs13061128
7. https://www.geomatyka.lasy.gov.pl/do-pobrania
8. Mosorov, V., Sankowski, D., Mazurkiewicz, Ł, Dyakowski, T.: The "best-correlated pixels" method for solid mass flow measurement using electrical capacitance tomography. Meas. Sci. Technol. **13**(12), 1810–1814 (2002)
9. Jaskulska, I., Jaskulski, D.: Winter wheat and spring barley canopies under strip-till one-pass technology. Agronomy **11**(3), 426 (2021). https://doi.org/10.3390/agromy11030426
10. Cabrera-bosquet, L., Molero, G., Stellacci, A.M., Bort, J., Nogués, S., Araus, J.L.: NDVI as a potential tool for predicting biomass, plant nitrogen content and growth in wheat genotypes subjected to different water and nitrogen conditions. C Cereal Res. Commun. **39**(1), 147–159 (2011). https://doi.org/10.1556/CRC.39.2011.1.15
11. Melillos, G., Hadjimitsisa, D.G.: Using Simple Ratio (SR) vegetation index to detect deep manmade infrastructures in Cyprus. Proc. SPIE **11418**, 114180E-E114181 (2020). https://doi.org/10.1117/12.2557893
12. Chehbouni, J., Qi, A., Huete, A.R., Kerr, Y.H., Sorooshian, S.: Modified soil adjusted vegetation index, remote sensing of environment **48**(2), 119–126 (1994).https://doi.org/10.1016/0034-4257(94)90134-1
13. Eng, L.S., Ismail, R., Hashim, W., Baharum, A.: The Use of VARI, GIL, and Vigreen formulas in detecting vegetation in aerial images. Int. J. Technol. **10**(7), 1385–1394 (2019). https://doi.org/10.14716/ijtech.v10i7.3275

Author Index

A
Adamczyk, Karolina, 227
Antosz, Katarzyna, 24

B
Balanou, Maria, 149
Bracke, Stefan, 124
Brüggemann, Dominik, 124
Bula, Karol, 161

G
Gevorkyan, Edwin, 113
Gola, Arkadiusz, 39
Golubenko, Aleksander, 93

H
Horník, Jakub, 202

K
Kania, Henryk, 189
Karkalos, Nikolaos E., 149
Karmiris-Obratański, Panagiotis, 149
Karwasz, Anna, 216
Kneifel, Jens, 124
Komarova, Hanna, 113
Korzeniewski, Bartosz, 161
Kowalski, Maciej, 138
Krajanowski-Kaleta, Jakub, 15
Kreibich, Viktor, 175, 202
Krol, Oleg, 93, 103
Kuchař, Jiří, 175, 202
Kudláček, Jan, 175, 189

L
Łożykowska, Natalia, 161

M
Markopoulos, Angelos P., 149
Michalski, Jakub, 1
Morozow, Dmitrij, 113
Mróz, Adrian, 15

N
Nerubatskyi, Volodymyr, 113

O
Oarcea, Alexandru, 50
Osiński, Filip, 216

P
Papazoglou, Emmanouil L., 149
Paśko, Łukasz, 24
Pizoń, Jakub, 39
Popişter, Florin, 50

R
Rucki, Miroslaw, 113
Rybarczyk, Dominik, 64, 78
Rząsa, Mariusz R., 227

S
Saternus, Mariola, 189
Sedláčková, Eva Michelle, 202
Siemiatkowski, Zbigniew, 113
Sokolov, Volodymyr, 93, 103
Stachowiak, Mateusz, 64
Stan, Sergiu-Dan, 50
Strek, Tomasz, 1
Svoboda, Jakub, 189
Świć, Antoni, 39

B. Gapiński et al. (Eds.): MANUFACTURING 2022, LNME, pp. 239–240, 2022.
https://doi.org/10.1007/978-3-031-00805-4

Szatkowski, Dawid, 78
Szymański, Maciej, 15

T
Tatíčková, Zuzana, 175
Trifan, Costan-Vlăduț, 50

U
Urbański, Jędrzej, 64

W
Wiecha, Mateusz, 227

Z
Zawadzki, Przemysław, 15, 138
Zoubek, Michal, 175, 189
Żywicki, Krzysztof, 15

The manufacturer's authorised representative in the EU is Springer Nature Customer Service Centre GmbH, Europaplatz 3, 69115 Heidelberg, Germany. If you have any concerns regarding our products, please contact ProductSafety@springernature.com

Printed and bound by CPI Group (UK) Ltd, Croydon, CR0 4YY
30/09/2026
02225320-0008